Energy and Geop(

MW01491615

The idea that energy shapes and is shaped by geopolitics is firmly rooted in the popular imagination – and not without reason. Very few countries have the means to secure their energy needs through locally available supplies; instead, enduring dependencies upon other countries have developed. Given energy's strategic significance, supply systems for fuels and electricity are now seamlessly interwoven with foreign policy and global politics.

Energy and Geopolitics enables students to enhance their understanding and sharpen their analytical skills with respect to the complex relations between energy supply, energy markets and international politics. Per Högselius guides us through the complexities of world energy and international energy relations, examining a wide spectrum of fossil fuels alongside nuclear and renewable energies. Uniquely, the book also shows how the geopolitics of energy is not merely a matter for the great powers and reveals how actors in the world's smaller nations are just as active in their quest for power and control.

Encouraging students to apply a number of central concepts and theoretical ideas to different energy sources within a multitude of geographical, political and historical contexts, this book will be a vital resource to students and scholars of geopolitics, energy security and international environmental policy and politics.

Per Högselius is Professor of History of Technology and International Relations at KTH Royal Institute of Technology, Sweden. His teaching and research centre on energy and natural resources in an international political context.

Energy and Geopolitics

Per Högselius

Routledge
Taylor & Francis Group

LONDON AND NEW YORK

from Routledge

First published 2019
by Routledge
2 Park Square, Milton Park, Abingdon, Oxon OX14 4RN

and by Routledge
52 Vanderbilt Avenue, New York, NY 10017

Routledge is an imprint of the Taylor & Francis Group, an informa business

British Library Cataloguing-in-Publication Data
A catalogue record for this book is available from the British Library

Library of Congress Cataloging-in-Publication Data
A catalog record has been requested for this book

ISBN: 978-1-138-03838-7 (hbk)
ISBN: 978-1-138-03839-4 (pbk)
ISBN: 978-1-315-17740-3 (ebk)

Typeset in Sabon
by codeMantra

Contents

Figures

Boxes

Acknowledgements

This book has two origins. The first is a graduate course, *Energy and Geopolitics,* that I developed together with my colleagues Arne Kaijser and Anna Åberg at KTH. First given in 2009, the course took inspiration both from our own research, which targeted various aspects of energy in transnational and historical perspectives, and from a number of current developments in the early twenty-first century relating to radical oil price hikes, Russian gas supply disruptions, transnational solar energy initiatives and so on. When looking for suitable course books, however, we found, to our surprise, that there was not much out there in terms of basic, comprehensive readings on the geopolitics of energy. In particular, it was difficult to find any books that combined experiences from the past – save a few iconic events such as the oil crises of the 1970s – with analyses of current energy affairs. There was also a heavy bias in the literature towards analyses of oil, at the expense of other energy sources. A few years later, having given the course a number of times and still being dissatisfied with its reading list, I decided that there was no use in complaining about the lack of suitable books – and so the idea was born to write my own book about the topic.

The other origin of the book is a research project, *Long-term Sustainability vs. Short-term Security.* Funded by the Swedish Research Council Formas, it gave me the chance to elaborate in depth on some of the key issues that we addressed in the Energy and Geopolitics course, especially in the risk domain. I had originally promised Formas to explore conflicting interpretations of and responses to risk in several different infrastructure systems. But as the project progressed, I found myself increasingly caught up with transnational energy infrastructures and how risk perceptions on radically different spatial and temporal scales interacted in the generation and restructuring of global energy supply. These considerations played an important role in shaping my writing.

I am indebted to my bright and always inspiring students, colleagues and collaborators in these undertakings. Apart from Arne and Anna, who contributed decisively to developing the Energy and Geopolitics course in the first place, I should also mention Dag Avango, Hanna Vikström and Suyash Jolly, who have helped me give the course and further develop it since 2012. Many other colleagues at the Division of History of Science, Technology and Environment at KTH have inspired me. The Stand Up for Energy strategic research programme has been another intellectual arena of great value, especially when it comes to bringing me closer to natural scientists and engineers, without whose knowledge it is often impossible to bring clarity to energy's geopolitical dimension.

The Tensions of Europe Network, which brings historians of technology of different breeds together on a European scale, has been a great source of inspiration, as always, while stimulating critical thinking on transnational energy issues. A one-year fellowship at the Chinese Academy of Sciences in Beijing (2013–2014) allowed me to discover the geopolitics of energy as it appears from the Middle Kingdom's horizon, as did my interaction with Chinese students in the 2015 and 2017 versions of the Renmin University International Summer School. In the final stages of the book project, I further profited from the activities of the Energy Materiality Study Group at the Hanse-Wissenschaftskolleg at Delmenhorst, Germany, organized by Margarita Balmaceda, while an invitation from the German Mining Museum in Bochum to give the keynote address at the King Coal conference held in winter 2018 forced me, at precisely the right moment, to rethink the geopolitics of coal – an otherwise often-neglected topic in energy studies.

I am grateful to the Swedish Research Council Formas and the Chinese Academy of Sciences for financial support and to Annabelle Harris at Routledge for encouraging me to take on the task of actually writing this book. Thanks also to four anonymous reviewers for their enthusiasm and constructive criticism on the original book proposal, to Susan Richter for improving the language and to Katherine Kay-Mouat for helping me out with the images.

I dedicate the book to Arne Kaijser, my great mentor and friend, who more than anybody else has inspired my thinking about energy in society.

1 Introduction

The idea that energy shapes and is shaped by geopolitics is today firmly rooted in the popular imagination. Two decades into the twenty-first century, it may even seem that the supply of energy in the world is inherently geopolitical, because it is virtually impossible to find a component in the global energy system that is not part, in one way or the other, of the earthly political drama. Conversely, it is almost impossible to think about modern geopolitics without making constant reference to energy – be it in the form of fossil fuels, nuclear power or the quest for renewables.

For most people, mentioning "energy" and "geopolitics" in the same sentence evokes memories of dramatic international events and crises such as the 1973 Arab oil embargo, Russia's alleged use of natural gas as a foreign policy tool or American visions of attaining "energy independence". Others may think of the radical disruption of energy supplies during the two World Wars, modern Iran's quest for access to uranium enrichment technology or European plans to tap into North African solar energy. But these developments, important as they may be, are just the tip of a huge iceberg. In the twenty-first century, the geopolitics of energy is everywhere. It is not merely something we find "out there"; it affects all of us as humans and interferes with our everyday lives.

Living in Sweden, I sometimes consider this when I wake up on a dark winter morning and turn on the lights in my apartment. When I turn on the lights, I become part of an electricity system that relies on huge turbines installed in distant Arctic rivers, from which electricity, through an internationally interlinked transmission system, is brought to my home in Stockholm; on nuclear power plants relying on uranium ore mined in Australia, transported across the oceans and enriched in France; and on offshore wind turbines erected in the Baltic Sea. I contemplate that the waterfalls in northern Sweden are able to generate electricity only because early twentieth-century industry and government leaders, fearing the potential consequences of Sweden's growing dependence on British coal for electricity production, thought it worthwhile to invest in this domestic energy source – Sweden's "white coal" – and because indigenous Sami protests against the physical alteration of the North's fragile environments were ignored in the process. As for Sweden's nuclear power plants, they are able to generate power only because the country's government has signed the international Non-Proliferation Treaty – a *sine qua non* for any country that wishes to source uranium through diplomatically approved channels of trade – and because the Australian government has decided that the far-reaching environmental destruction that inevitably results from uranium mining is tolerable.

My colleagues in continental Europe may at the same time be scrambling their breakfast eggs on gas-fired stoves. The gas that heats their pans is piped into their houses through an intricate distribution system, the main supply sources for which are buried in sandstone under the Siberian tundra, deep down in the Saharan desert and beneath the salty depths of the North Sea. The gas flows into their stoves only because political and business leaders, in the midst of the Cold War and radical decolonization processes in Algeria, Libya and elsewhere, managed to negotiate long-term contracts for imports of these faraway gas sources. In fact, the available volumes of imported natural gas in Europe are so large that Europeans use it not only for cooking purposes but also for heating their water and their houses in winter. Industries and electric power plants have also taken advantage of the foreign gas streams. European food supply likewise depends on non-European gas because it is the main raw material for producing nitrogenous fertilizers. But for how long will the gas keep flowing? After all, natural gas is a fossil – and thus finite – fuel. In most countries, this resource is already close to exhaustion. In the Netherlands, once a major gas producer, production is now rapidly being phased out due to a tragic course of depletion which has recently given rise to a series of strange earthquakes in the Groningen region. But most European analysts agree that natural gas is direly needed to manage the transition to renewables. And this gas will have to come from ever more remote fields.

Meanwhile, my friends at the Chinese Academy of Sciences in Beijing, being seven hours ahead of me, are perhaps returning home early from work so as to avoid the huge traffic jams that have been plaguing the Middle Kingdom's capital city for many years now. Beijing has opened a dozen or so new subway lines since the early 2000s, but it is individualized rather than public transport, more than anything else, that has captured the imagination of the Chinese urban middle class. The dream of a personal car is pervasive, and at the time of this writing, a petrol-driven vehicle remains the self-evident choice for Beijing's car buyers. The number of cars in China has more than tripled from 64 million in 2008 to 194 million in 2016,[1] and there has been a corresponding growth in filling stations. The latter are owned by Chinese state-controlled oil companies, which face the delicate task of procuring all the crude oil that is necessary for my academic colleagues to be able to regularly fill up their vehicles. The oil companies do so not only by buying oil on the open market, but above all through "strategic partnerships" with the oil industries of Kazakhstan, Angola, the Democratic Republic of the Congo (DRC) and elsewhere. There, the companies'

Figure 1.1 Road traffic in modern Beijing. The dream of a personal car is pervasive in China, contributing decisively to the country's rapidly growing oil-import dependence. Photo: Per Högselius.

aggressive investments are interpreted by critics as just another form of colonialism and, in Central Asia, bricks in a "New Great Game" – a reference to the late nineteenth-century struggle for control over this region between Britain, Russia and China.

So the geopolitics of energy is closer to us than we might think. And it shapes our world views – sometimes negatively, sometimes positively. It confirms or challenges our perceptions about whether everything is going to hell or, by contrast, if a new bright era in the history of humankind – or of our own country – is emerging. To some, energy's marriage with geopolitics has generated a fatal "race for what's left", an escalating struggle for increasingly scarce yet vital resources that is bound to end in a global, apocalyptic war.[2] Others are convinced that a different kind of apocalypse – that of global warming – can similarly be traced back to energy in its geopolitical context, because it is only through the movement of vast volumes of fossil fuels across the globe that their massive combustion and hence large-scale carbon emissions become possible. Another popular understanding is that poverty, injustices and civil wars in much of the "Third World" can be explained only in terms of fabulous – but abused – resource riches in countries such as Iraq and the DRC.

The optimists point to the geopolitics of energy as a creative force in national and international politics. In Vladimir Putin's Russia, for example, global might in oil and gas has become a surrogate for Russia's lost superpower status – and, by extension, a key constituent of Russian national identity. While foreign observers hardly subscribe to the idea that this is something positive, the dream of mobilizing Russian resource wealth to build a prosperous, modern Russia is undoubtedly still alive and kicking. French national identity, for its part, is linked to nuclear energy, which contributes to the country's literal and metaphorical radiance in the world, while Norway's "responsible" management of its resource wealth goes hand in hand with this small nation's understanding of itself as a "humanitarian superpower". For Azerbaijan, energy – and the 2011 victory in the Eurovision Song Contest – is what makes the nation "European". In European Union (EU) politics, meanwhile, energy now seems to offer a welcome opportunity to forge closer intra-European ties in an age of looming refugee crises, economic quarrels between north and south over budgets and debts, and political separatism pointing towards disintegration. And what, if not a natural gas pipeline from Russia to the Korean peninsula, could become the trigger for political reunification between North and South?

What is energy?

Both "energy" and "geopolitics" are notoriously slippery terms. "Energy" is closely linked to the history of physics, but it has since become a popular concept in many other sciences as well, including the social sciences. Standard dictionary definitions teach us that energy is "the capacity for doing work" – but this does not bring us far when it comes to energy in the social and political realm. Physicists tell us that energy cannot be created or destroyed. Yet we would be unable to navigate the realm of energy in society without referring to the "production", "consumption" and "waste" of energy. The idea that some forms of energy are "renewable" sounds like witchcraft to the physicist's ear, yet it is part and parcel of modern energy policymaking. And while the laws of thermodynamics assure us that energy is always "conserved" in one way or the other, the quest for energy conservation and storage remains a daunting challenge.

In the social and (geo)political context, energy sources do not exist in any objective sense; they "become" energy through various social, political and economic activities.[3] Take oil: the oil industry experienced its breakthrough during the second half of the nineteenth century. But oil was not an unknown substance in pre-industrial times. Ancient philosophers like Aristotle and Pliny the Elder wrote about the black liquid that came seeping up from the ground. In Greece and Babylonia, applications were found for the material in medicine, pharmacy and agriculture. It was also used in refined form as a literal – rather than, as in our own era, metaphorical – weapon. Petroleum was also a well-known substance in China. In North America, Paleo-Indians collected oil on a fairly large scale for use in their religious rituals. Centuries later, in the mid-1840s, things started to change. There was now a great demand for lighting appliances – in homes but also in factories. Industrialization was coming, and factory owners wanted to be able to operate their expensive machines even when it was dark outside, especially in wintertime. Firewood and vegetable oils, the traditional forms of light, would not do for this purpose. Whale oil was emerging as a popular alternative. Then somebody came up with the idea of making use of coal to produce a kind of lighting oil. In the 1850s, it was discovered that this kind of oil could also be obtained through the distillation of petroleum. The resulting liquid was called kerosene. It became immensely popular, and a new energy source was born – after thousands of years of non-energy uses of "rock oil".[4]

Natural gas, for its part, was long regarded as a nasty by-product of oil extraction. It was a dangerous thing, liable to cause lethal explosions and jeopardize industrial installations and business profits. To deal with this problem, producers started "flaring" gas – in other words, it was burnt in the air. Such gas fires may still be seen in many oil-producing regions,

Figure 1.2 Oil gusher in Baku, 1883. After several millennia of non-energy uses of petroleum, "rock oil" emerged as a large-scale energy source in the second half of the nineteenth century. Photo: Kochne W. & Co./Swedish National Museum of Science and Technology.

indicating that natural gas is regarded less as something useful than as a waste product. Subsequently, oil industrialists came up with the idea of using gas as an energy source in its own right. For this purpose, they started building pipelines to connect oil and gas fields with industrial and urban regions. There, the gas was put to work for cooking, heating and various industrial purposes. And so natural gas became an energy resource.[5]

Equally telling is the story of how uranium came to be viewed as an energy resource. In Roman times, uranium oxides were found useful for colouring purposes in the ceramic industry. In the Middle Ages, uranium ore began to be extracted on a larger scale from silver mines near Jáchymov, in what is now the Czech Republic, from where it was brought to nearby glass-making industries. In the late nineteenth century, steel industrialists took an interest in uranium and started to use it as an alloying metal. Elsewhere, however, uranium, like natural gas, was typically perceived as an unwanted waste product. Gold miners in South Africa were constantly annoyed by it, as it made people ill. Then, during World War II, the bomb-makers arrived, seeing in uranium the potential for a new weapon of hitherto unimaginable force. Only in the years around 1950 did the idea seriously emerge of uranium as an energy source. Today it is a major, and controversial, source of electricity in many countries.[6]

In this sense, over centuries humans have not merely "discovered" various energy sources; they have "invented" them. It is crucial to be aware of this aspect because the "invention" process is by no means complete. In present-day debates about bioenergy, for example, it is hardly clear which trees and crops may be regarded as energy – especially in view of the many competing uses these materials serve, notably in the food domain. And what kind of material substances will "become energy" tomorrow? The waves rolling in from the ocean are one of the most recent targets for engineers in search of new things that can become exploited. Some scientists and engineers even believe that seawater in and of itself may be turned into an energy source.

Some scholars refer to invention in this sense as processes of "social construction", indicating that it is in the social realm that much of the global energy system is "built". Whether or not wave energy will become a success, for example, is not merely a matter of technological successes and failures but also depends crucially on whether a critical mass of support for the idea can be mobilized among financiers, politicians, regulators and, not least, the general public. As we will see, not only energy as such is "invented" or "socially constructed"; different kinds of risks and uncertainties – and how they are weighed against each other in debates and decision-making processes – are subject to similar forces, as are the much discussed "energy weapons".[7]

Humans have also come up with the idea that energy sources can become internationally traded as "commodities". While "energy" refers mainly to something materially useful, the notion of coal, oil, gas and other energy sources as commodities is distinctly economic. It has to do with dollars, euros and yuans rather than with joules or watts or a hot bath. The process of *commodification*, through which energy sources are assigned a market value (or an "exchange value", as Marxists prefer to call it), is key to anyone interested in energy and geopolitics.[8] Without it, most fossil fuels and uranium ores would most probably have remained in the ground; they would not have "become" energy. Few (if any) human agents would have been interested in extracting and refining them on a large scale because there would not have been any revenue stream to make up for the huge investments needed to extract the fuel and bring it to consumers. The importance of energy sources being traded as commodities is most evident in market economies. But even in supposedly non-market economies, such as the former Soviet Union and its satellite states in Central and Eastern Europe, fuels were sold and bought on a grand scale, both domestically and internationally. Today, the global energy system is inextricably linked with international commodity markets in a way that decisively shapes the geopolitics of energy.

This book analyses energy sources both in their material forms – that is, the substances that are perceived of as constituting "energy" – and as tradable commodities. We will explore various fossil fuels – coal, oil, gas – along with nuclear energy, renewables and everything in between. Furthermore, we will not only study "crude" forms of energy like petroleum or uranium ore; we will also study the extraction and production of primary energy sources as taking place within larger systems, in which refined forms of energy – like petrol or electricity – play key roles. A key argument of the book is that we cannot properly grasp energy and geopolitics without taking into account the *totality* of global energy, viewing different energy sources and energy forms as constituents of a single whole. It is not that useful, for example, to study only oil, because the geopolitics of oil is seamlessly interwoven with that of other energy sources. As we shall see, the need for such a holistic perspective becomes even more pressing when we move into the realm of renewables and the effects of the ongoing "energy transition" on geopolitical power relationships.

What is geopolitics?

"Geopolitics" is just as tricky a concept as "energy". In its most generic sense, geopolitics is about the interaction between geographical factors, politics and international relations. The term may be felt to be value-laden, and historically it has commonly been associated with the "realism school" in international relations theory. The term remains tainted by the fact that a range of dictators, from Adolf Hitler to Augusto Pinochet, have been fond of it. Klaus Dodds, one of the best authors on the subject, notes that even today, geopolitics, more often than not, "is used by politically conservative journalists and pundits ... as a shorthand term, intended to convey a robust attitude towards political action".[9] "Spheres of interest", "rising powers", "heartlands", "living space" and similar terms belong to the favourite vocabulary of these analysts. However, alternative conceptualizations are both possible and desirable. In particular, there is a need for the geopolitical analyst to take into account not only the high-political arenas inhabited by national presidents, foreign ministers and national security advisors. A much wider range of actors has to be considered. And in terms of places, it is clearly not enough to consider what happens inside the "White House" or the "Kremlin"; we need to broaden our geopolitical view and learn to paint with a much wider – and more colourful – geographical palette.

In energy, this becomes particularly obvious. We may listen to what a Xi Jinping or a Vladimir Putin have to say – both often refer to energy when commenting on international political developments – but this does not take us far when seeking to unravel the political dynamics of international energy. We need to move "up" to transnational corporations and international organizations of various kinds, but also "down" to regional and local actors. We need to take into account the perspectives of businesses small and large, of independent entrepreneurs, professional associations, trade unions and environmental movements – all of which have historically seen a range of opportunities and risks in international energy and which have acted accordingly. The Texas Railroad Commission in the United States, the Bavarian regional government in Germany, the Lithuanian anti-nuclear movement, the Southern African Development Community (SADC), Austrian steelmakers and striking British coal miners are just some of the actors that historically have been decisive in shaping the geopolitics of energy. Needless to say, we also need to follow the activities of scientists and engineers, who more than once have upset the global energy arena with their bold ideas, dreams and inventions. Yesterday the geopolitics of energy was transformed by the invention of nuclear fission, catalytic cracking and high-pressure gas pipelines. Today it is being transformed by electric vehicles, solar energy and second-generation biofuels – as well as by shale gas, carbon capture and storage and deep-sea drilling for oil and gas.

Where is the geopolitics of energy "made"? While political activities in capital cities like Tehran, Brussels or Beijing clearly deserve attention, global energy also produces its own geography, with key nodes that are more often than not off the map in the regular political geography. This geography includes places such as Cerréjon in Colombia (home to a huge coal mine of immense significance for European energy supply), the Yamal-Nenets Autonomous Region in north-western Siberia (which rests on some of the world's largest natural gas deposits), the ocean bed off the Brazilian coast (where some of the most breathtaking activities to capture the world's remaining oil resources are under way) and the city-state of Singapore (which sits on the strategic Malacca Strait while also being home to the world's largest palm oil biodiesel plant). Anonymous localities "in between" – that is, between the production and consumption hotspots of the global energy

Figure 1.3 Gulf of Mexico oil spill recovery, June 2010. Depletion of fossil fuels in easily accessible regions is transforming the global energy production geography, pushing oil explorers to remote locations like this one. The shift is accompanied by new environmental risks, as evidenced by the *Deepwater Horizon* disaster in April 2010. Here, ships and drilling rigs surround the *Discoverer Enterprise* (upper centre) as it continues to recover oil from the destroyed drill site in the Gulf of Mexico. Over the next three months, some 4.9 million barrels of oil welled up from the seabed. Photo: US Coast Guard/Science Photo Library.

geography – also play prominent roles. Baumgarten, to take just one example, is little more than a village in Lower Austria, but on the maps of European and Russian gas companies, it looms large like a major capital city because of its key role as a European "hub" in the East-West gas trade. And if there is a geographical centre of gravity in world oil, it is the Strait of Hormuz at the entry to the Persian Gulf – a waterway that does not attract many tourists, but which is both larger and more iconic than London, Paris or New York for those who make a living from petroleum.

Like energy, the geographical and geopolitical features of the world are best understood as "made" or "invented". This may be so in very concrete ways, as when forests are cleared to make space for agriculture, when towns and cities expand into rural areas or when the world's wet geography is remade by humans through the straightening of rivers, the construction of artificial canals and dams and the dredging of shallow sea lanes.[10] This physical transformation of the natural world is intimately linked to continuous transformations in human geography. But we also construct geographical and geopolitical features in our imagination. The geographical vocabulary seems to require terms such as "North" and "South", "East and West", "centre" and "periphery" and so on. But when we point to

Germany as being located in the "centre" of Europe, what do we actually mean? Lithuanians argue that the "real" centre of Europe – following accepted natural-geographical definitions of "Europe" – is located near a village called Purnuškės, not far from Vilnius. And while North Europeans think of Portugal and Greece as distinctly southern nations, the global discourse about "North-South relations" places them in the northern "camp". Clearly, the notions of "centre" and "periphery" and "north" and "south" do not always relate to scientifically measurable or objective features and facts; instead, they reflect social, economic and political power relations. But we do need them in our analysis.

The making and remaking of global energy supply continuously shifts our ideas about what constitutes "centres" and "peripheries" in the geopolitics of energy. The centre of oil production moved from Pennsylvania to California in the late nineteenth century and then to Texas and Oklahoma in the early twentieth. Russian oil moved from Baku in the Caucasus to the Volga-Urals region and then, in the Cold War era, to north-western Siberia – also known as the "Third Baku" (while the original Baku became an oil periphery). From the 1850s right up to the 1940s, the United States dominated global oil, but then the centre of gravity shifted to the Middle East. For centuries, the centre of global coal production was located in northern China, where industries and households made this fuel their main source of energy as early as the Song Dynasty. Later on, Britain took over this position, emerging as the unchallenged world coal centre, with much of Europe becoming dependent on British coal supplies. China became a coal periphery, only to re-emerge as coal's heartland in the early twenty-first century. Britain, meanwhile, has now totally phased out domestic coal mining. Today we see a new, intense struggle for centrality in renewable energy geographies. Some believe that the world's deserts, given their huge solar energy potential, will emerge as the main centres in the age of renewables. Others point to Bolivia as the future "Saudi Arabia of lithium", the main mineral needed for the batteries in electrical vehicles, or to countries such as Germany and Japan, which can claim technological leadership in a number of areas related to renewables.

Outline of the book

Energy and Geopolitics is structured along six main chapters. Chapter 2 provides the basis for the rest of the book by developing a *sociotechnical systems perspective* on global energy. The chapter takes inspiration from infrastructure research in science and technology studies (STS) and the history of technology and from an understanding of geopolitics as the politics of space, distance, connections and materiality in a wide sense. The prefix "geo" means "earth" in Latin, but the geopolitics of energy is not only about the energy that is in and on the Earth – fossil fuels, uranium, forests, winds and waves. It is about long-distance movements of energy; it is about the tanker ships, the pipelines and the transmission lines needed for this trade to be feasible and about the places where energy in its various forms is produced, refined, stored and consumed. Energy and geopolitics, it is argued, are characterized by "messy complexity", and a systems approach helps us to bring analytical order and discern the most important patterns in this complex – and exciting! – world.

Chapter 3 turns to the mesmerizing variety of *actors* that in one way or another play significant roles in the geopolitics of energy. It starts out by discussing the role of private enterprise as a powerful actor category, for which the lure of energy sources as tradable commodities takes centre stage. It then turns to the heterogeneity of the state as an actor, whose interests are less clear-cut, and its quest for control over energy in the transnational context. The lives of state-owned energy companies, as a special category of public actors, are also scrutinized. Moreover, state actors at subnational – regional and local – levels need to be taken into account. The focus then shifts to a number of additional actor categories, such as trade unions, environmental non-governmental organizations (NGOs), scientists and media – all of which have contributed decisively to shaping the geopolitics of energy over the years.

Chapter 4 draws on the systems perspective and the actor-based approach developed in the preceding two chapters to unravel the dynamics of *energy dependence*. The chapter starts out by outlining the main patterns of (inter)dependence in global energy and how these have changed over time, often in unexpected ways. The systems perspective helps us to deconstruct simplistic statistical indicators of dependence, emphasizing that actors in different countries depend on each other not primarily for energy supplies as such but rather for the international systems through which the energy moves and transforms. Such a view generates new, unexpected geographies of energy dependence. A further theme of central importance concerns the underlying motivations of actors when it comes to engaging – or not engaging – in international energy relations. We discern, in particular, how (strongly subjective) perceptions of opportunities and risks have shaped global energy supply in fateful ways.

Chapter 5 turns to actors' ambitions to "do something" about what they consider to be problematic dependencies and vulnerabilities in the international arena. Two overarching strategies are at focus: *reducing* dependence by, for example, increasing reliance on domestically produced energy; and *coping* with dependence. The chapter discusses a heterogeneous set of sub-strategies and "tools" for coping with dependence, from strategic stockpiling programmes and attempts to spread risks by diversifying fuel supplies (in terms of geography and fuel mix) to violent military action and various forms of "energy diplomacy". Of particular importance here is how state actors "securitize" foreign energy – that is, by defining it as being of "strategic" interest and essential for national survival.

Chapter 6 starts out by reminding us that since energy systems are not merely technical, but rather sociotechnical constructs, their purposes cannot be defined objectively. Different actors will always see energy systems as having different purposes. On this basis, the chapter discusses how foreign policy actors have come to define energy as having a purpose far beyond energy supply as such. We go through the different ways that energy systems can be – and have been – manipulated for foreign policy purposes. While manipulations of energy flows and energy prices are central here, the chapter also stresses that system-building processes and public discourses about energy have often been subject to similar manipulations. In addition to this, we look at the socially constructed nature of "energy weapons" while emphasizing that not only state agencies but a range of non-state actors, too, have sought to make use of energy as a foreign policy tool.

Chapter 7, finally, follows up on the preceding chapter but shifts the focus from negative to positive uses of energy in the foreign policy context. *Energy transnationalism*, the central concept here, refers to attempts by actors to mobilize energy for the purpose of strengthening international political stability, preventing wars and stimulating transnational cooperation. This idea, which builds on philosophical concepts originally developed in the decades around 1800, is omnipresent in today's globalized world – and it challenges the popular idea that energy and geopolitics are primarily a harsh, warlike "battle" for global supremacy and control.

Exercises

- Can you identify any physical materials and/or natural phenomena, other than the ones mentioned in this chapter, that are likely to "become energy" in the future? Can you identify any materials or phenomena that have *ceased* to be viewed as energy?
- Have all energy sources been commodified? Can you think of any physical material or natural phenomenon that, worldwide or in a particular region, is "only" an energy source and not a commodity?
- Compile a list of at least ten geographical places that, in recent years, have figured prominently in international energy affairs.

Notes

1 See, for example, the statistical portal Statista, www.statista.com/statistics/285306/number-of-car-owners-in-china/.
2 Klare, *The Race for What's Left*.
3 Zimmermann, *World Resources and Industries*, 15.
4 For a detailed account of this early history of oil, see, for example, Black, *Crude Reality*.
5 Högselius et al., "Natural Gas in Cold War Europe".
6 Emsley, *Nature's Building Blocks*, 476–482; Hecht, "Colonial Networks of Power", 153–154.
7 The concept of social construction is used in a variety of academic fields. In this book, I take inspiration especially from science and technology studies (STS) and the history of technology, with seminal works such as Bijker et al., *The*

Social Construction of Technological Systems, but also from the Copenhagen
School in International Relations and various works by geographers.

8 Two useful accounts of commodification as an historical process (in the context of natural resources) are Black, *Crude Reality,* and Cronon, *Nature's
Metropolis.*
9 Dodds, *Geopolitics,* 3–4.
10 For example, Högselius et al., *Europe's Infrastructure Transition,* Chapters
7–9.

Further reading

Black, Brian. *Crude Reality: Petroleum in World History.* Lanham, MD: Rowman &
Littlefield, 2012.
Dodds, Klaus. *Geopolitics: A Very Short Introduction.* Oxford: Oxford University
Press, 2007.
Klare, Michael T. *The Race for What's Left: The Global Scramble for the World's
Last Resources.* New York: Metropolitan, 2012.

Bibliography

Bijker, Wiebe, Thomas P. Hughes and Trevor Pinch, eds. *The Social Construction
of Technological Systems: New Directions in the Sociology and History of Tech-
nology.* Cambridge, MA: MIT Press, 1987.
Black, Brian. *Crude Reality: Petroleum in World History.* Lanham, MD: Rowman &
Littlefield, 2012.
Cronon, William. *Nature's Metropolis: Chicago and the Great West.* New York:
Norton, 1991.
Dodds, Klaus. *Geopolitics: A Very Short Introduction.* Oxford: Oxford University
Press, 2007.
Emsley, John. *Nature's Building Blocks: An A-Z Guide to the Elements.* Oxford:
Oxford University Press, 2001.
Hecht, Gabrielle. "Colonial Networks of Power: The Far Reaches of Systems". *An-
nales Historiques de l'électricité* 2 (2004): 147–157.
Högselius, Per, Anna Åberg and Arne Kaijser. "Natural Gas in Cold War Europe:
The Making of a Critical Infrastructure". In *The Making of Europe's Critical
Infrastructure: Common Connections and Shared Vulnerabilities,* edited by Per
Högselius, Anique Hommels, Arne Kaijser and Erik van der Vleuten, 27–61.
Basingstoke and New York: Palgrave Macmillan, 2013.
Högselius, Per, Arne Kaijser and Erik van der Vleuten. *Europe's Infrastructure Tran-
sition: Economy, War, Nature.* Basingstoke and New York: Palgrave Macmillan,
2016.
Klare, Michael T. *The Race for What's Left: The Global Scramble for the World's
Last Resources.* New York: Metropolitan, 2012.
Zimmermann, Erich. *World Resources and Industries: A Functional Appraisal of
the Availability of Agricultural and Industrial Materials.* New York: Harper &
Brothers, 1951.

2   International energy systems

"Messy complexity"[1] is perhaps the most adequate term to describe the reality of world energy in the twenty-first century. At times it may seem like a hopeless task to try and make sense of what's going on and why. Take natural gas, for example. In order to understand its global career in recent years, we have to take into account not only much-debated themes such as the spectacular rise of the US shale gas industry or the debate about Russia's alleged uses of natural gas as an "energy weapon". We have to look into new technical advances in fields such as gas liquefaction technology, subsea pipe-laying, gas-turbine making and the development of synthetic alternatives to natural gas. It also becomes necessary to analyse a range of environmental issues, from the role of natural gas in causing – and combatting! – climate change to the severe air pollution that plagues many large cities in the world and the state-led attempts in countries such as China to cope with it by making use of natural gas. We should trace the impact of sudden unexpected events such as the 2011 Fukushima nuclear disaster, which not only reshaped nuclear energy's future worldwide but also upset East Asian gas markets in a way that nobody could have foreseen. We have to study new infrastructural projects such as the politically contested gas pipeline from the Caspian Sea to Southern Europe and, in another part of the world, the efforts to widen the Panama Canal, through which US shale gas can now be exported more or less profitably to Asia. Furthermore, we need to familiarize ourselves with European Union (EU) regulatory reforms and gas directives, with export and import subsidies in Asia and the Middle East and with the rise and fall of protectionist policies in the United States. We have to keep track of mergers, acquisitions and foreign investment in the gas industry. We have to keep an eye on international quarrels about how to politically govern the world's oceans. And we have to look into conflicts over extraction and transmission of natural gas in politically and ethnically sensitive regions such as the Altai Autonomous Republic in Russia and the Sinai desert in Egypt, home to Bedouin minority groups.

But this multitude of dimensions should not hinder us from trying to understand what's happening. This chapter suggests that a *systemic perspective* is useful in taking on the challenge of analysing energy in its global, geopolitical context. The word "system" may signal different things; it is used by a wide range of scientific communities, from computer experts and evolutionary biologists to economists and sociologists. But they have one thing in common: they take a broad, holistic perspective, seeking to understand the totality of things and how everything is interconnected with everything else. There is, undoubtedly, something sympathetic about this: the refusal to turn the smaller bits and pieces into the main units of analysis; as these Buddhists of the academic world argue, it is more important to fit everything together into one and the same understanding of the world. While challenging to adopt, a systemic perspective holds the potential to add invaluable new layers of meaning to the study of energy and geopolitics.

What is an energy system?

What, precisely, is an energy system? For our purposes, it suffices to think of it as a set of technical and non-technical components – plus the links between them – that jointly enable energy in one or the other form to be delivered to end users. Depending on how we define the *boundaries* of such systems, the actual unit of analysis may differ. While it is possible, in principle, to define a "planetary" (or "global") energy system, encompassing the totality of the world's energy production and consumption, in many cases it may be more useful to conceptualize energy supply as taking place within a set of more narrowly defined (sub-)systems. For example, we may want to single out one or the other energy source and analyse the dynamics and evolution of, for example, the oil supply system, the nuclear energy system or the biodiesel system. Alternatively, it may appear useful to study nationally, regionally or locally delimited (sub-)systems. However, these sub-systems are rarely closed entities; they communicate with their environments, and they influence each other in a variety of ways. Thus, one of the key challenges for the student of global energy is precisely to come to grips with the interaction between different sub-systems. For example, we may want to scrutinize the relations between two national energy systems, or the interaction between a fossil fuel system and a renewable energy system.

Importantly, the claim that the circulation of energy can be conceptualized as taking place within a system does not necessarily mean that there is a centrally placed operator or "dispatcher" somewhere who "controls" global energy flows. Energy systems, especially in the international context, are not necessarily centralized constructs; more often they are decentralized or have several centres of gravity, whereby a variety of actors usually compete for power and influence.[2] This becomes particularly clear if we accept that energy systems are not merely technical entities; as we will see, it is much more useful to view energy systems as *sociotechnical* in nature. This is to say that they comprise not only oil barrels, nuclear power plants, high-voltage transmission lines, wind turbines and street lights but also a multitude of *people* who plan, build, manage, operate, regulate and use the systems. Analysing the system's human actors and how they relate to its technical constituents is clearly a necessary exercise for anyone who wishes to understand (geo)political power structures in global energy supply.[3]

Primary and secondary energy systems – and their entanglement

It is useful to distinguish between *primary* and *secondary energy systems*. Primary energy systems revolve around the extraction, processing and use of *primary energy sources* such as firewood, coal, crude oil, natural gas, uranium or palm oil – materials that occur in nature and which humans decide to make use of as fuels. There are also a range of non-fuel primary energy sources, such as sunshine, winds, flowing water and (animal and human) muscle power. Secondary energy systems, as the term indicates, are centred on *secondary energy sources*. These do not exist in nature but have to be materially created on the basis of one or more primary energy sources. Today, electricity is the most important secondary energy source by far. Other examples of secondary energy sources of great importance include water-based heating and cooling systems and systems for the distribution of manufactured gases (such as coal-based gases or biogas).

The *entanglement* between primary and secondary energy systems increases the overall complexity in world energy supply – with far-reaching consequences for the geopolitics of energy. For example, if we want to assess the vulnerability of urban electricity supply in, say, Hamburg, it is clearly not enough to merely study Hamburg's local electricity network. We will, first of all, have to track down that network's role in northwestern Germany's regional electricity system, and the regional system's interconnections with foreign electricity systems in the wider European network. Then, we need to acknowledge that the European electricity system, being a secondary energy system, is entangled with a range of primary energy systems, the far reaches of which force us to spotlight a range of overseas resource peripheries. We will then find that electricity users in Hamburg depend on the smooth functioning of gas transmission systems that feed natural gas from northwestern Siberia, from the Saharan desert and from deposits buried under the North Sea into European gas-fired power plants; on the smooth operation of large coal mines in Colombia, South Africa, Australia and other remote places; on uranium being extracted from mines in Niger's desert lands, in the Kazakh steppes of Central Asia, and in northern Saskatchewan; and even (albeit to a lesser extent) on palm oil harvested from large plantations in Southeast Asia. Within a not all too distant future, Hamburg's electricity security may further come to depend on huge solar energy parks in North Africa, the Middle East and Central Asia.

As long as the lights stay on, the actual extent of these entanglements remains invisible. Only in times of crises can we get a true sense of how primary and secondary energy systems are interconnected into a wider whole. World War I was the first major crisis that, especially in Europe, highlighted the vulnerability of the intricate entanglements between systems of various kinds. Cities went dark as gas and electricity supplies were disrupted, and a range of traditional energy sources, such as firewood, vegetable oil and peat, made unexpected comebacks.[4] World War II produced an even more severe energy crisis, forcing countries to take extreme measures to protect their electricity and manufactured gas supplies at a time of shortage of primary energy sources. Today, the massive blackouts that from time to time plague countries such as India and South Africa are also closely related to the entanglement between the electricity system and one or the other dysfunctional primary energy system.

Energy and ICT

Energy's far-reaching integration with spatially extensive information and communication systems constitutes a further dimension of entanglement. This is of particular importance in the case of *grid-based energy systems* such as electricity, natural gas and district heating. Electrical engineers were fast to recognize the potential of information and communications technologies (ICTs) to strengthen control over electricity flows. Initially the control systems were local. However, when local electrical grids gave way to larger, regionally interconnected electricity systems during the first decades of the twentieth century, the need for coordination grew. In this situation, the public telephone network came to play a vital role for coordinating power production. This entanglement between electricity and telephony created new dependencies and vulnerabilities.

BOX 2.1 A BLACKOUT IN OSLO, 1930

One day in 1930, the lights suddenly went out in Oslo, Norway's capital city. It first appeared that the event was caused by excess electricity production. But, as it turned out, the crisis had a different underlying reason: there was a problem with the telephone lines. The crisis report revealed that the machine operator at one of Norway's hydroelectric power plants had been using the station's only telephone to discuss a political matter with a comrade in the local Labour Party – precisely at the time when the phone was needed to communicate urgent system updates! The dispatcher at the Power Pool in Oslo observed the overload in the system. However, since the telephone line was busy, he was unable to get through with his order to halt a generator.[5]

Today, electricity's digitalization and entanglement with data communications networks are seen to offer opportunities for building new, "smarter" electricity grids. But just like in the early twentieth century, the heavy reliance on ICT for system control is also seen to present new vulnerabilities. The possibility of "cyberattacks" directed towards vital electricity, gas and district heating systems causes substantial headaches among the world's system operators. Energy security is increasingly being blended with cybersecurity.[6]

Further entanglements: water, minerals, food

Equally pervasive is energy's far-reaching entanglement with water, minerals and food. Water is of critical importance for energy supply. Most obviously, the world's rivers have, to a very large extent, been mobilized for electricity production. The construction of hydroelectric facilities in different parts of the world has often been highly controversial, as these have far-reaching effects on other potential uses of the river's water. Fisheries, irrigation, shipping, log driving and fresh water supply are some of the activities that may be endangered through the construction of hydropower plants – although in many cases hydropower also opens up opportunities to strengthen other activities. The history of hydroelectricity is full of conflicts, many of which are played out on the geopolitical arena. Some recent cases include the damming of rivers such as the Tigris, the Brahmaputra and the Nile.

Water is also needed for the cooling of large-scale nuclear and fossil-fuelled power plants. A typical nuclear power plant needs a water flow amounting to around 50 cubic metres per second to operate safely. The water needs to continue flowing at all times; otherwise, the nuclear reactor will start to heat up in a dangerous way and a core meltdown is around the corner. For this reason, basically all nuclear power plants in the world are located in the immediate vicinity of very large natural bodies of water – seas, lakes and rivers. In some international river basins, notably those of the Rhine and the Danube in Europe, nuclear energy's immense thirst for water forced nuclear power plant builders in different countries to coordinate their plans at an early stage. Germany, France, Switzerland and the Netherlands all built nuclear power plants along the Rhine and its tributaries. The existence of a well-established international organization for the protection of the Rhine helped the parties involved to cooperate and deal with conflicts in this process. Yet nuclear energy in the Rhine Valley, as well as along the Danube, remains highly controversial in European politics.[7]

The most recent controversy concerning energy's entanglement with water is the fear that groundwater aquifers will be polluted as a result of "unconventional" production of oil and gas. It is mainly this fear that has put an end to European dreams of emulating the US shale gas revolution, and it remains a highly controversial issue in the United States itself. In countries

Figure 2.1 The Fukushima nuclear power plant in Japan in 2007. A majority of the world's nuclear power plants are located in the immediate vicinity of the sea. This is because they are in need of massive water supplies. This also makes them highly vulnerable to flooding, as demonstrated most tragically by the disaster at Fukushima in March 2011. Photo: Tokyo Electric Power Co. (TEPCO) / IAEA.

such as Poland and Ukraine, shale gas has at times been viewed as a pathway to energy independence from Russia, but not everyone agrees that it is worth risking clean water supplies for the sake of enhanced energy supply security.[8]

Minerals are also of critical importance for energy systems. If water is needed mainly for the regular operation of energy systems, minerals are needed mainly in the formative and growth phases in the system's career (see below for a discussion of development phases in the evolution of energy systems). In electricity systems, copper has always played a critically important role, and a large share of the world's copper production now ends up in electricity grids. In the nuclear industry, zirconium has been one of the most critically important minerals. More recently, it has become clear that the transition from fossil fuels to renewables depends on ample access to further "critical metals". Lithium, cobalt and rare earth elements are the ones that have mostly been at stake here so far. Lithium and cobalt currently play a key role in the production of batteries for electrical vehicles as well as for electricity storage facilities (crucial for electricity systems to be able to handle large amounts of intermittent solar and wind power). Rare earths, comprising elements such as neodymium and dysprosium, are of corresponding importance for the powerful magnets that are used in wind turbines, while platinum group metals are key to the production of high-efficiency solar cells. The availability – and market price – of these minerals influences the competitiveness of renewable energy sources *vis-à-vis* fossil fuels and nuclear energy. As a result, international conflicts over access to the desired minerals have become part and parcel of energy and geopolitics, and renewable energy policies in the Western world have become deeply entangled with large-scale mining activities in places such as Chile and Argentina (for lithium), the Democratic Republic of the Congo (for cobalt) and Inner Mongolia in China (for rare earths).[9]

Food and energy supply systems have been closely interlinked for millennia. Some energy historians even insist on counting food as an energy source in its own right, especially when the pre-industrial era is concerned, since food (and fodder) has always formed the basis for human and animal muscle power.[10] Environmental historian Joachim Radkau, for example, has pointed to the "oat limits" of overland transport in pre-industrial Europe; before the rise of coal and railways, overland transport depended on the availability of fodder for horses in the form of oats, the supply of which was naturally limited by the available agricultural land.[11]

The transition to fossil fuels generated new entanglements between energy and food. With the invention and spread of the Haber-Bosch process, through which nitrogen and hydrogen are synthesized to produce ammonia, fossil fuels started to be used for the production of artificial fertilizers. This allowed for a radical intensification of agriculture and scaling up of global food production to unprecedented levels.[12] Today, natural gas is the most important source used in the production of nitrogenous fertilizers, and much of the world's natural gas is used precisely for this purpose. Disruptions of natural gas imports might thus have adverse effects on a nation's food supply.

A more recent, and highly controversial, example of energy-food interlinkages is the enormous investments, in the twenty-first century, in literally *growing* energy. Bioenergy has a long history, but the spectacular rise in biofuel feedstock production in recent years is unique. When it comes to crops such as corn, rapeseed and oil palms, energy and food uses compete directly with each other, with global market prices playing an important role in determining whether a harvest will be used for fuel or rather for food. In the early twenty-first century, when fuel prices loomed high, this competition was one of the factors leading to a severe food crisis in the developing world.[13]

Competition between energy systems

Energy systems based on different primary and secondary energy sources do not operate in isolation from each other. We have already noted that secondary energy systems are entangled with various primary energy systems.

At the same time, two or more systems may compete with each other in their struggle for dominance. For example, fossil fuel systems now compete fiercely with renewable energy systems in the electricity market, while electricity and biofuels compete with various oil- and gas-based substances in the transportation sector.[14]

Actors representing different energy systems use a range of strategies in seeking to win the competitive struggle. First, they may try to convince customers that "their" energy system offers *technical opportunities* which their competitors are unable to offer, or that their energy system is more *environmentally friendly*. Electricity companies early on mobilized such arguments in their competition with, for example, town gas and kerosene suppliers in late nineteenth-century lighting markets. In the post-World War II era, oil suppliers competed with coal companies by pointing to the greater convenience of handling oil compared to coal. They also stressed the environmental benefits for consumers in the heating market of switching to oil, since oil burning did not produce the same level of air pollution as coal burning. Today, electricity and natural gas actors point to their energy systems as superior to both coal and oil, citing both technical performance factors and environmental aspects.

Second, different energy systems compete against each other on *price*. For example, while the transition from coal to oil in some energy markets was attractive from a technical and an environmental point of view, it was only when oil prices fell below those of coal in the 1950s and 1960s that users found it really attractive to switch from one primary energy system to the other. And in some countries with exceptionally low electricity prices – such as Norway and Sweden – electricity has been able to compete successfully with both coal and oil in parts of the heating market. Pricing is obviously also a critical factor in determining the pace of the current transition from fossil fuels to renewable energy systems.

Third, energy companies and other actors may *lobby governments and state agencies*, seeking to win their support when it comes to issues such as taxation, subsidies and overall energy policy decisions. Historically, governments have often taken decisions that directly influenced the competition between energy systems. Geopolitical factors have played prominent roles in this context. Notably, governments tend to favour domestically available energy sources over imported ones – even if the domestic ones are more expensive or have poorer technical and environmental performance. Poland's huge coal industry is a case in point. Polish coal is favoured over imported natural gas, even though this makes neither economic nor environmental sense; the main reason for still insisting on coal is that the country's natural gas, to an overwhelming extent, comes from Russia. India's heavy reliance on (mainly domestic) coal can also be interpreted as the result of geopolitical considerations. In this case, access to alternative energy sources such as uranium or natural gas is linked to complex international issues such as India's problematic relations with neighbouring Pakistan (a potentially important transit country for natural gas on its way to India from Iran and Turkmenistan) and the West's refusal, before the conclusion of the 2005 Indo-US nuclear deal, to give India (which has not signed the Non-Proliferation Treaty) access to international uranium markets. As a result, coal has been the most competitive energy system in India.

The British government's decision to switch from domestic coal to imported oil for fuelling its navy in the early twentieth century offers an interesting counter-example; here, Winston Churchill opted to sacrifice a domestic energy system for a highly internationalized one – for the sake of oil's much higher technical performance in the operation of big warships. Another case is the Swedish state's decision to phase out domestic uranium mining, in favour of a nuclear fuel supply regime centring on imports; pricing and environmental issues weighed more heavily than self-sufficiency considerations.[15]

Today, political decisions continue to play key roles in shaping the competition between energy systems, not least in the fierce struggle between "old" systems linked to fossil fuels and nuclear energy and "new" systems tied to renewable energy sources. However, in most countries, direct governmental intervention has now become relatively rare; it has been replaced by indirect measures such as adjustments in energy taxes, state subsidies, import tariffs and environmental regulations.

Figure 2.2 Competition between energy systems in the age of renewables. Greenpeace activists and Munduruku install solar panels at Dace Watpu village in the heart of the Amazon. The Munduruku people have inhabited the Sawré Muybu Indigenous Land for generations. The Brazilian government plans to build a series of hydroelectric dams in the Tapajós River basin. Fearing that this would severely threaten their way of life, the Munduruku have joined forces with Western environmental activists, arguing that solar energy makes more sense than hydropower. Photo: Otávio Almeida/ Greenpeace.

The extent to which energy systems actually compete with each other has historically varied very much over time, in response to changes in performance (especially in terms of numerous technical breakthroughs), prices and political trends. Coal, for example, enjoyed a remarkable career during the early days of industrialization. By the early twentieth century, it was being used by industrial enterprises for smelters, industrial steam engines, locomotives and ships as well as for space heating. Coal emerged as a universal fuel, amenable to nearly any purpose. Later on, however, it was out-competed by other energy systems for most of its earlier uses. In the developed world, today coal is used almost exclusively for electricity production ("steam coal") and for iron ore smelting ("coking coal").[16] In many developing countries, it is still used for space heating and for cooking, but the overall trend is the same everywhere: coal is being squeezed out from several of its historical core markets. The trend is the same for oil. At one point this fuel seemed to be on its way to becoming a truly universal fuel. It successfully competed with a range of other energy systems because it was cheap, easy to handle and environmentally friendly (i.e. in relation to coal, its main competitor in non-transportation markets). After the oil crises of the 1970s, however, many countries, especially those that did not sit on any oil deposits of their own, began phasing out oil in sectors such as electricity production and heat supply. The only field where it was found difficult or even impossible to replace oil with something else was in transportation. When electrical vehicles, biofuels and other non-oil technologies now start competing in earnest with gasoline, diesel and jet fuel, we will witness the oil industry's struggle for its last remaining markets. If, or rather when, oil is squeezed out from the transportation market, there may be no significant market left for oil, save for its use in the petrochemical industry.

By contrast, electricity, as a secondary energy source, is competing ever more successfully with alternative (primary and secondary) energy systems. Especially in the ongoing transition to renewables and in the development of electrical vehicles, electricity assumes an ever-greater role. Electricity is the only energy carrier that can ensure the integration, over vast spaces, of growing volumes of wind, solar, wave and tidal energy. Furthermore, it is increasingly expected that electricity will take over in the transportation sector, too – both with regard to the vehicles themselves and in the construction of a new, electrical road infrastructure. These trends are bound to make electricity ever more central to the geopolitics of energy.

Basic activities

An important task for the systems-oriented analyst is to try and make sense of the diversity of *activities* that take place in energy systems. Historically, when locally available wood played the main role in world energy supply, the activities were, in most localities, limited to collecting firewood from nearby sources and then burning it. In modern energy systems, the picture is much more complex, and a much more diverse range of activities take place.

Most obviously, as soon as energy needs cannot be met from locally available sources, *transportation* becomes a critically important activity. There are two types of transport infrastructures relevant in this context. The first is the *general transport infrastructure*, which takes the form of roads, railways, rivers, canals and sea lanes. Even back in pre-industrial times, firewood was sometimes moved from one place to another through log-driving on rivers and streams and, in some cases, by shipping along coasts or across the sea. As a rule, however, the technical difficulties and high costs of such transport limited the utility of the long-distance fuel trade. In the early modern period, artificial waterways were increasingly enlisted. The seventeenth-century Dutch Golden Age, for example, is difficult to imagine without abundant and cheap access to peat and the unique canal network built for transporting this fuel. Later on, canals began to be used for coal transports, too, whereby coal companies often took active part in new canal construction. In England, for example, Duke Francis Egerton built the famous Bridgewater Canal (1761) to ship coal from his estates to Manchester's important coal market. Many canals in France, both before and after the French Revolution, were likewise heavily used for coal deliveries. The coming of rail further widened the opportunities. Many railways, ranging from England's pioneering Stockton-Darlington line to colonial projects like Germany's Shantung Railway in China, operated primarily for coal shipments. Since railway companies were themselves major coal users, railway construction and the expansion of coal movements mutually supported each other.[17]

While the general transport infrastructure has continued to play an extremely important role, it has been increasingly complemented by *dedicated energy infrastructures*. These are networks that do not rely on general transport, and which cannot be used for any other purpose than energy transmission and distribution. Over time, dedicated energy infrastructures have come to shape the internationalization of world energy in decisive ways. Oil and gas tankers, pipelines and electricity transmission lines are important examples. The first dedicated energy infrastructure to emerge historically was the pipe networks built for the distribution of town gas. London's Gas Light and Coke Company built the world's first urban gas system, inaugurated in 1814.[18] Starting in the 1870s and 1880s, the town gas networks were accompanied by another dedicated energy infrastructure: electricity networks. Initially, both gas and electricity networks were local constructs. In the twentieth century, however, they grew to become large, interconnected systems. Today, gas and electricity networks often span entire continents. Dedicated energy infrastructures have also emerged in the oil industry, with oil tankers and pipelines playing particularly prominent roles. Even coal is sometimes moved over long distances through pipelines.[19]

Figure 2.3 A man looks at the world's biggest Liquefied Natural Gas (LNG) tanker, DUHAIL, as she crosses through the Suez Canal on 1 April 2008. The Qatari tanker, which was built to transfer LNG from Qatar to Europe and the United States, is on her first trip ever from Qatar to Spain. Natural gas is notoriously difficult to transport, and liquefaction is a prerequisite for overseas shipments. Photo: Reuters/TT.

Second, modern energy sources such as coal, oil, uranium or solar energy generally cannot be used directly; they first have to undergo *conversion* and *refinement* of some kind. In the wooden energy age, some of the wood was refined into charcoal, mainly for use in the steel industry, but most wood was burnt directly. In the modern era, the situation is very different. Oil refineries constitute the best-known modern conversion facilities; without them, petroleum would be a useless fuel. In nuclear power, refining plays an equally important role, although the relevant processes are here referred to as "conversion" (of milled uranium ore into uranium hexafluoride) and "enrichment" (through which the concentration of the fissile uranium isotope U-235 is increased). Uranium enrichment, in particular, has always been a geopolitically sensitive activity. The great powers, notably the United States, have consistently tried to prevent smaller nations and developing countries from acquiring enrichment technology, the argument being that such technology can potentially be used to develop not only civilian nuclear power but also nuclear weapons. Iran's attempts to acquire uranium enrichment capability, and the sanctions devised by the West to prevent this, are a case in point.

In the case of coal and natural gas, refining is less central. In many cases these fuels can be used more or less directly. However, gradual depletion of the best deposits continuously raises the importance of refining activities here, too. Hence, gas treatment plants, for example, do play a role in numerous gas-producing regions. Coal refining, for its part, is most important in the case of low-value coals such as lignite. The development of lignite refining technologies has historically played a key role in making lignite-based energy systems competitive in relation to more advanced energy systems.[20]

Third, no deposit of non-renewable energy lasts forever. Hence, if the system is to continue functioning, let alone further expanding, there is a constant need to compensate for depletion by identifying replacement deposits. As a result, *geological surveying, prospecting and exploration* become crucially important activities to sustain non-renewable energy systems. Some discoveries of vast new fossil fuel deposits, especially oil and gas, have become iconic events in global energy history. The discovery of the Groningen gas field in the Netherlands in 1959 is a case in point; that of the Prudhoe Bay oil field in Alaska in 1968 another. In the early days of fossil fuel extraction, prospecting and exploration were actively encouraged by state actors, who hoped that new domestic fuel finds would strengthen energy autarky. Over time, however, most nations have faced depletion of their own resources, and so they have become dependent on the discovery of new deposits abroad. Today, prospecting and exploration is a strongly internationalized activity. A growing number of state agencies and private companies compete ever more fiercely with each other for the world's remaining fuel reserves, which are concentrated in an ever-smaller number of countries.[21]

Fourth, many modern energy sources give rise to problems at the "back end" of the system, as the materials at hand produce not only energy but also a variety of *undesired by-products* that may cause health and environmental problems and hence have to be disposed of in one way or the other. This has often been linked to issues of a geopolitical nature. Spent nuclear fuel and radioactive waste are a case in point, especially when it comes to the reprocessing of spent nuclear fuel. Starting in the 1970s, the spent fuel reprocessing facilities in France, Britain and Russia became sites of international importance as a range of smaller nuclearized countries set out to ship their spent fuel for reprocessing there. The shipments quickly became controversial, not only in view of the fact that reprocessing of spent fuel produces plutonium – which can be used in the construction of nuclear weapons – but also because of the ethical dimension of export as a way of "coping" with domestic waste problems. Today exports of spent fuel are declining, and when they do take place, the nuclear waste that is produced through reprocessing is usually sent back to the country of origin. In the meantime, direct disposal of spent nuclear fuel – an alternative strategy to reprocessing – has also become subject to extensive international debate. So far, the conventional wisdom holds that every country should build its own geological repository. However, the International Atomic Energy Agency (IAEA) has long advocated the construction of large international disposal facilities where many countries could jointly bury their nuclear waste. The argument is that this will be geologically and economically more rational and efficient.[22]

Spent nuclear fuel aside, the most burning "back-end" issue in the twenty-first century is clearly that of the carbon dioxide emissions that inevitably result from burning fossil fuels and bioenergy. While the main line of debate here has focused, of course, on *reducing* these emissions, a range of (geo-)engineering solutions are also being promoted for *coping* with carbon. Carbon capture and storage (CCS) is the most well-known. Currently, it appears uncertain as to whether CCS will ever see a breakthrough, but if it does, a whole new transport infrastructure will have to be added to fossil fuel and bioenergy systems.[23]

Fifth and finally, modern energy systems often face problems because energy is not produced and delivered in the same "rhythm" as it is consumed. Hence, there is a need for *storage facilities* of various kinds. To this day, energy storage remains one of the most difficult challenges for system-builders to cope with. Oil storage facilities are the most well-known. Gas storage facilities are of equal importance, but since they usually take the form of underground facilities they are much less visible to the public. Gas, together with electricity, offers enormous technical and economic challenges as far as storage is concerned, and yet this activity is crucial for guaranteeing the smooth operation of the corresponding energy systems. In the case of electricity, renewable energy system-builders are now struggling to develop effective mechanisms for storing electricity from intermittent energy sources.[24]

Over time, energy storage has also emerged as one of the main tools to cope with other kinds of fluctuations in the flow of energy. In particular, it has become a standard safeguard against politically motivated supply disruptions and price shocks in the international fuel trade. Hence, the International Energy Agency's (IEA's) requirement that its member states store a volume of oil equivalent to at least 90 days of consumption has nothing to do with the day-to-day fluctuations in international oil flows. Instead, it reflects the fear of a geopolitically triggered oil crisis with massive supply disruptions.

Exactly what constitutes a "storage facility" is subject to interpretation. For example, oil and gas storage as a vulnerability management tool has historically often taken the form of "shut-in" production: oil and gas fields that, under normal circumstances, are not used to the extent that they might be, but which may be drawn upon in case of crisis and so compensate for any removal of other supplies. During World War II, for example, US shut-in supplies of oil played a key role in ensuring that the Allied forces in Europe were adequately supplied with oil. The arrangement was successful and was revived again and again in the post-war era: first in connection with the Korean War and Iranian nationalization crisis of 1950–1953, then during the Suez crisis of 1956–1957 and finally during the Six-Day War in 1967. By 1970, however, domestic US oil supplies had become scarce, and shut-in production was no longer available. This largely explains why the Western world was so vulnerable to the Arab oil embargo of 1973.[25]

The formation and evolution of energy systems

Energy systems are never permanent constructs. They come and go, and they change a lot over time. Stability is always temporary at best. But how do energy systems actually come into being, and what are the forces that determine their patterns of evolution?

Energy systems do not come into existence through magic, or by mere chance. They are the result of bold visions, engineering ingenuity, financial risk-taking, difficult entrepreneurial decisions, tough negotiations – and a lot of hard work. Often one or a few centrally placed individuals play decisive roles in creating the system and shaping its subsequent evolution. Borrowing a term from the Large Technical Systems (LTS) approach to sociotechnical systems analysis, originally developed by American historian of technology Thomas P. Hughes, we may refer to these key persons as *system-builders*.[26] These may be technically oriented innovators, but more often they are passionate business leaders or centrally placed governmental actors who have the necessary ability, mandate and connections to bring about major infrastructural projects, turning diffuse and often controversial visions into material reality. Having a talent in viewing the system in its totality, spotting the links between its diverse technical, political and economic components, the successful system-builder identifies "reverse salients" in the form of weak components and links, and turns these – analytically and discursively – into "critical problems" that must be solved for the system to come about and expand along desired lines. Well-known historical examples of energy system-builders include John D. Rockefeller, who more than anybody else shaped the emerging oil supply system in the late nineteenth century, and Thomas Alva Edison, the inventor who played the main part in creating America's first electricity systems.

Hughes distinguishes among four phases in a system's development.[27] The first is what we may refer to as the *formative phase*, in which the system is invented and first developed.[28] It is a phase in which historians of technology take great interest, and where visions of things that do not yet have a clear future figure prominently. Systems which are in the formative phase do not look impressive at all in statistical reports because they are still small and may not yet have experienced their breakthrough. Indeed, it is often very difficult to prophesize about whether or not a system that is in its formative phase will ever "make it" and become a large, commercially successful system. For example, the nuclear fusion energy system has been in a formative phase for half a century, and nobody knows whether it will ever become commercial. If we look at science and engineering journals, systems in their formative phase are all over the place, being passionately discussed in terms of technical and societal visions, the chances that they will be able to out-compete already existing systems, and the risks that everything will fail. It is easy to both underestimate and overestimate a system that is in its formative phase, because a range of technical, economic, institutional and political uncertainties may decisively influence its viability.

The *second phase* in an energy system's career centres on *technology transfer* – in other words, attempts by system-builders in a different geographical setting to emulate – or build a different version of – an original system. Although in some cases a new energy system may emerge in parallel in two or more geographical settings, it is more common that it first emerges in one place, after which the news of its success inspires actors elsewhere in the world to "copy" it. For example, Edison's original electricity system, built in New York in the years around 1880, inspired system-builders in London and Berlin to build their own versions of that system. And in the case of the early oil industry, Rockefeller's system, comprising both key technologies and (highly controversial) business methods, inspired Russian oil industrialists to construct a similar system. Technology transfer in this sense is often linked to geopolitics. As mentioned, the governments of the Western great powers have long sought to prevent, by diplomatic and other means, other countries from emulating their nuclear fuel systems, because these comprise technologies of military significance. In a similar vein, during the Cold War, several US governments sought to prevent the Soviet Union from building successful versions of American oil and natural gas systems. It is highly likely that we will see further geopolitical controversies stemming from technology transfer in the context of the ongoing transition to renewables; the recent trade war between China and the EU in solar cells gives a hint of what's at stake.

The *third phase* in an energy system's development is characterized by its *material growth and geographical expansion*. This phase features complex integration processes in which early local systems become interlinked through long-distance transportation systems. The main challenge for system-builders here becomes to balance the growth of different key activities. The oil system's historical growth, for example, demanded not only the identification of new, large oilfields and scaled-up production of crude from these, but also new innovations in oil transport (such as supertankers), massive refinery construction all over the world, new oil-storage facilities, and, last but not least, aggressive efforts by state-owned and private oil companies to conquer – in competition with each other and with other energy systems – new markets.

In the *fourth and final phase*, the system has matured. Growth is less pronounced, but the system continues to develop. It has now reached a high level of *momentum*, which makes it difficult for it to "change direction".[29] Enormous amounts of capital have now been invested in the system, countless people and careers are bound up with it, and overall the system is so integrated with key political, economic and social functions that it becomes difficult to imagine a world without it. Today's oil system is an obvious example. More and more people nowadays agree that it would be better both for the planet and for human civilization if fossil fuels were phased out altogether. Yet the internal dynamics of the oil system, with its excessive momentum, points to further growth and development of oil production and consumption as the "natural" way forward. According to the internal logic of the oil system, depletion of the more accessible resources is not a major problem; it is countered by the development of new technologies such as hydraulic fracturing and deep-sea drilling, and by the expansion of prospecting and exploration to regions like the Arctic Ocean and war-torn countries in Africa and elsewhere. Carbon emissions, for their part, are countered through other innovations, such as CCS. As technological ingenuity becomes more and more important for further expanding the system, higher education and research are further expanded; specialized petroleum universities in many countries, from Russia and the United States to India and China, nowadays accept ever larger numbers of undergraduate and graduate students whose careers will be bound up with oil. And when all else fails, criminal activities may be a way out, as in the automotive industry's recent attempts to hide the truth about actual emission levels from diesel-fuelled vehicles. The momentum of the oil system "forces" it to expand in the same direction as before.

When analysing energy and geopolitics, we need to be aware of the phase a given system is in at a given point in time. For example, while North American, Japanese and European oil systems are now clearly in the momentum phase, the oil systems of many developing countries are still in the growth phase. Europe's challenge nowadays is basically to maintain its current level of supply (and European demand is actually decreasing). By contrast, China and India are in a situation where oil demand is projected to increase at an astonishing pace, and their oil companies face the immense challenge of accessing ever larger volumes of crude. This is the main logic driving Chinese and Indian investments in foreign oil production, especially in Africa, but also in Central Asia, Latin America and other places.

International system-building: economy vs. politics

System-building in transnational contexts is particularly challenging. Apart from differences in standards, regulations, political traditions and business cultures, system-builders may have to cope with problematic legacies of bi- and multilateral relations from the past. Xenophobia and historical traumas of various kinds have more than once put an end to international system-building attempts in energy. Crucially, system-builders setting out to cooperate with "the other" will have to accept that they cannot take control over the system-building process to the same degree as they may be used to doing in local and national environments. As a result, the construction of transnational energy systems becomes a process with several centres of gravity. The accumulation of critical levels of trust between the involved actors emerges as one of the keys to success, especially in the system's formative phase. Hence, if there is a suspicion that the partner with whom the system is to be jointly built will later use that system to exert political pressure – using it as a political "weapon" – it is unlikely that the system will actually be built. There are numerous examples of would-be transnational energy systems that have failed to materialize due to lack of trust. In many other cases, however, transnational system-building has become a highly dynamic process, with effective coalitions of system-builders turning the apparent problems of cross-border tensions and disparities into opportunities for mutually beneficial development and growth. Without such success stories of transnational cooperation, world energy supply would today still be a mainly local affair.[30]

Seeking to understand the logic that drives international system-building, we may first of all remind ourselves that the world's energy sources are unequally distributed. Some countries and regions are richly endowed with domestic energy sources, while others are not.

BOX 2.2 THE UNEQUAL DISTRIBUTION OF THE WORLD'S ENERGY RESOURCES

Where are the world's most important fuel deposits located?

As of December 2016, the countries with the *largest oil reserves* were Venezuela (with 17.3% of global reserves), Saudi Arabia (15.6%), Canada (10.0%), Iran (9.3%), Iraq (9.0%), Russia (6.4%), Kuwait (5.9%), the United Arab Emirates (5.7%), the United States (2.8%), Libya (2.8%), Nigeria (2.2%), Kazakhstan (1.8%), Qatar (1.5%) and China (1.5%). 71.5% of the reserves were under the control of the Organization of the Petroleum Exporting Countries (OPEC).

Actual *oil production* was dominated (as of 2016) by Saudi Arabia (13.4% of global production), the United States (also 13.4%), Russia (12.2%), Iran (5.0%), Iraq (4.8%), Canada (4.8%), the United Arab Emirates (4.4%), China (4.3%), Kuwait (3.4%), Brazil (2.8%), Mexico (2.7%), Venezuela (2.6%), Norway (2.2%), Qatar (2.1%) and Angola (2.0%). The OPEC member states produced 42.7% of the total.

The largest countries in terms of *natural gas reserves* were Iran (18% of proved global reserves), Russia (17.3%), Qatar (13.0%), Turkmenistan (9.4%), the United States (4.7%), Saudi Arabia (4.5%), the United Arab Emirates (3.3%), Venezuela (3.1%), China (2.9%), Nigeria (2.8%) and Algeria (2.4%).

Actual *gas production* was dominated by the United States (21.3% of global production), Russia (16.3%), Iran (5.7%), Qatar (5.1%), Canada (4.3%), China (3.9%), Norway (3.3%), Saudi Arabia (3.1%), Australia (2.6%), Algeria (2.6%), Malaysia (2.1%) and Indonesia (2.0%).

The *largest coal reserves* were held by the United States (22.1% of global reserves), China (21.4%), Russia (14.1%), Australia (12.7%),

India (8.3%), Germany (3.2%), Ukraine (3.0%), Kazakhstan (2.2%) and Indonesia (2.2%).

Actual *coal production* was dominated by China (46.1% of global production!), the United States (10.0%), Australia (8.2%), India (7.9%), Indonesia (7.0%), Russia (5.3%), South Africa (3.9%) and Colombia (1.7%).

Uranium production was led by Kazakhstan (39.4% of global production!), Canada (22.5%), Australia (10.1%), Namibia (5.9%), Niger (5.6%), Russia (4.8%), Uzbekistan (3.9%), China (2.6%) and the United States (1.8%).

The world's largest *biofuels producers* were the United States (43.5% of global production!), Brazil (22.5%), Germany (3.9%), Argentina (3.4%), Indonesia (3.0%), France (2.7%), China (2.5%), Thailand (2.0%) and the Netherlands (2.0%).

Hydroelectricity, in terms of consumption rather than production, was led by China (28.9% of the world total!), Canada (9.7%), Brazil (9.6%), the United States (6.5%), Russia (4.6%), Norway (3.6%) and India (3.2%).

The main consumers of *other renewables* were China (20.5% of global consumption), the United States (20.0%), Germany (9.0%), Brazil (4.5%), Japan (4.5%), the United Kingdom (4.2), India (3.9%), Spain (3.7%) and Italy (3.6%).[31]

For the "have-nots", the good news is that the construction of international energy systems allows them to access energy from the "haves". A country such as Sweden, for example, does not produce any fossil fuels at all, nor any uranium ore. Yet it boasts of a highly successful steel industry which thrives thanks to imported coal and coke; the country also has millions of car owners and truck drivers who are able to fill up their petrol- and diesel-driven vehicles at the country's 3300 petrol stations, and airplanes that are able to take off and land in a multitude of Swedish airports – all based on imported oil. Natural gas, supplied by a single pipeline from Denmark, heats towns and supplies industries with process heat in Sweden's south and southwest, and several large nuclear power plants, which run on imported uranium, produce copious amounts of electricity for the benefit of Sweden's households and highly competitive energy-intensive industries.[32] Most European countries, along with the most advanced nations in East and Southeast Asia, are in similar situations, although, of course, the degree of dependence on imported fuels varies. The most typical country is perhaps one that has access to some domestic energy (Sweden thus benefits from its unusually large hydropower potential) while being forced to import most of its remaining energy needs.

Hence, bringing in energy from foreign sources becomes a critical activity for energy system-builders in most nations; in the twenty-first century, it is extremely rare that a locality can fully cover its energy needs from locally available sources. Even the more richly endowed countries are in no way independent of foreign energy supplies. The geopolitics of energy, importantly, is very much about defining the conditions under which the necessary long-distance movements of fuel and electricity may take place. In particular, it is about assessing the technical, economic and political feasibility of different transport routes. We may discern two competing logics here, the first of which takes the *natural geography* of fuel endowments as its point of departure, while the second centres on the *political geography* of the world.

The natural geography logic states that it is economically rational for fuels to move from localities where they are abundant and can be produced cheaply, to places where there is a high demand for the same fuels. A century ago, for example, at which time coal had become the dominant energy source in Europe, it was seen rational for coal supplies to move from key coal-mining regions in Britain and Germany to larger cities and industrial centres elsewhere in Europe. System-builders in both the exporting and the importing countries understood this and set out to jointly create a corresponding trade regime. In the post-war era, when large deposits of natural gas were discovered in the Netherlands, Libya, Algeria and the Soviet Union, the natural geography logic similarly pointed to the rationality of moving gas from these countries to Western Europe's industrial heartlands. Again, system-builders responded to this logic by creating a vast infrastructure of pipelines through which the gas could be moved to where it was needed. Australia's energy exports are another case in point; being sparsely populated but richly endowed with fossil fuels and uranium, the natural geography logic pointed to the advantages of Australian fuel shipments to overseas markets. In the case of coal, such shipments remained uncompetitive before World War I, but thanks to radically reduced freight costs in the decades that followed, Australian coal could soon be sold with a profit on markets as far away as Europe. Today Australia forms a cornerstone in the world energy system, being a key supplier of coal as well as of uranium and, increasingly, natural gas.

The political geography logic, however, challenges such "natural" trade patterns by pointing to the need for boycotting and punishing "unfriendly" or "unreliable" nations, or for prioritizing domestic users over customers abroad – even though this might be economically irrational. Before World War I, for example, Germany exported a lot of coal from its Silesian mines, in the southeast of the country. The trade made clear economic sense. After World War II, however, by which time the mines had become Polish, exports of coal from the very same deposits became highly controversial. Among other things, Britain and the United States, being worried about the political integrity of capitalist Western Europe, tried to prevent countries such as Sweden and Denmark, which were located in reasonable geographical proximity of the mines, from accessing Polish coal.[33] In a similar way, the Soviet Union's energy exports to Western Europe were repeatedly challenged on political grounds, even though most of these exports seemed to make very good economic sense. Another example is the cut-off, following the Korean War, of previously voluminous coal movements from the northern part of the Korean peninsula to the south. Today, much of the international trade in biofuels and biofuel feedstocks – especially from the global south to the global north – is prevented from following a "natural" trade pattern by a range of politically motivated tariffs and regulations in the EU and elsewhere, some of which are designed to protect EU-internal biofuel producers, and others to stimulate the environmentally and socially responsible cultivation of energy crops in the global south.[34]

As a rule, both the natural and the political geography logics have their supporters and opponents. Some system-builders consider political and security issues to be more important than economic and technical factors. Others may think it is the other way around. The point here is that the outcome of the competitive struggle between the two logics – and their respective champions – is never given. In Cold War Europe, the political geography logic celebrated several victories as existing electricity links between capitalist and communist countries were actually cut, while numerous visions of new interconnections across the Iron Curtain failed to materialize. But in natural gas, the natural geography logic won the battle, with vast amounts of fuel being brought in from the Soviet Union and a vast pipeline network constructed for this purpose – in spite of the ideological and military antagonism between East and West.[35] The natural geography logic has also been victorious in the case of cross-border gas system-building between Egypt and Israel, in spite of the fact that the political geography logic tells us that these two countries should not cooperate. The mismatch between the two geographical logics is also evident, for example, in the energy relations between Sudan and South Sudan, or between India and Pakistan – all of them countries that would seem to have a lot to gain, in economic terms, from cooperation in energy, but where the political geography logic tells them that cooperation should be avoided. As a result, these sites are among the most exciting to explore, analytically speaking, in present-day energy geopolitics.

All in all, the evolving geography of world energy can be conceptualized as an ongoing struggle between political and natural geographical logics in international system-building. However, the momentum that builds up in international energy systems over time complicates the picture. Once an international system has matured, it will appear extremely attractive to stick with, in keeping with the natural geography logic. Abandoning the system will usually appear out of the question for system-builders. On the contrary, system-builders will try to develop "their" system further. Conversely, it becomes extremely difficult – and expensive – to radically alter the system. A mature energy system with a high level of momentum almost becomes part of the natural environment, being superimposed on an existing geography of seas, rivers, forests and mountains. Like other large infrastructures, it becomes a "secondary nature" of the modern world.

But what happens to such mature systems when the political geography changes? Not only energy systems change over time. Countries also rise and fall, and political maps are redrawn. Take the collapse of the Soviet Union in 1991, for example, which Russian President Vladimir Putin famously labelled "the worst geopolitical disaster of the twentieth century". In Soviet times, an interconnected Soviet electricity system had been created, spanning multiple republics. The northwestern part of this system had originally been designed as a tightly integrated system that served both natural and political geography logics. It exploited the available primary energy sources in the region, for the benefit of the region as a whole, and it was also in line with central Soviet political ambitions to unite several different Soviet republics – Russia, Belarus and the three Baltic republics of Estonia, Latvia and Lithuania. The collapse of the Soviet Union and the independence of the three Baltic countries in 1991 suddenly meant that the integrated electricity system, which now had become an international one, was no longer desirable from a political point of view in the Baltics. Delinking from Russia and Belarus, however, did not seem to make economic sense at all, both because of the overall design of the system, which was economically favourable to all who were part of it, and because Estonia and Lithuania could now earn valuable export revenues by exporting electricity to Russia and Belarus. The momentum of the system made it appear irrational – even impossible! – to divide the system into several national constituents. As a result, 25 years after the collapse of the Soviet Union, the electricity systems of the three Baltic countries were still synchronously interlinked with those of Russia and Belarus – although in the meantime the former had become both EU and North Atlantic Treaty Organization (NATO) members.[36] A similar story could be told about the fate of the Central Asian electricity system or the integrated Transcaucasian natural gas grid following communism's collapse, or about the joint oil system of what used to be Sudan (mentioned above), now split into two independent nations. All in all, the momentum of energy systems often makes them more long-lived than the polities that were responsible for their construction in the first place.

Exercises

- Pick a town, city or region and try to find out how its local energy supply is entangled with other regions, domestic and foreign. In your analysis, make sure to consider the entanglement between primary and secondary energy systems.
- Select a national or international energy system and sketch its emergence and evolution over time. In which phase is the system now? To what extent is it characterized by a high level of momentum?

- Identify at least one cross-border coalition of system-builders (other than the ones mentioned in this chapter) that has played a role in forging new cross-border energy links.

Notes

1 Cf. Hughes, *Rescuing Prometheus,* 195–197.
2 Högselius, *Red Gas,* Chapter 1.
3 I here take inspiration from scholars such as Thomas P. Hughes and Arne Kaijser, whose works are listed in the bibliography.
4 Cordovil, "De-electrifying the History of Street Lighting"; Del Curto and Landi, "Gas-Light in Italy".
5 Thue, "Connections, Criticality, and Complexity", 223.
6 See further Högselius et al., *The Making of Europe's Critical Infrastructure.*
7 The author is currently heading a major research project targeting the entanglement between nuclear energy and water, NUCLEARWATERS: Putting Water at the Centre of Nuclear Energy History, funded by the European Research Council (ERC).
8 For example, Johnson and Boersma, "Energy (In)security in Poland".
9 For example, Vikström et al., "Lithium Availability and Future Production Outlooks".
10 E.g. Kander et al., *Power to the People.*
11 Radkau, *Wood.*
12 See, for example, Quader's insightful study of how natural gas system-building went hand in hand with the agricultural "Green Revolution" in Bangladesh. Quader, "Natural Gas and the Fertilizer Industry".
13 This is further discussed in Chapters 5 and 7.
14 Here I draw mainly on Kaijser et al., *Changing Direction,* 89–105.
15 Yergin, *The Prize,* 11ff; Hultgren and Olsson, "Uranium Recovery in Sweden".
16 See, for example, Smil, *Energy at the Crossroads,* Chapter 1.
17 For the Dutch peat transports, see, for example, De Vries, *The European Economy in an Age of Crisis,* 165f. For the role of canals in European coal transports, see the national case studies in Kunz and Armstrong, *Inland Navigation.*
18 Tomory, "Building the First Gas Network".
19 See, for example, Wasp, "Progress with Coal Slurry Pipelines".
20 For an in-depth account of the Bulgarian case, see Tchalakov et al., "Bulgarian Power Relations".
21 See, in particular, Klare, *Rising Powers, Shrinking Planet.*
22 Högselius, "Spent Nuclear Fuel Policies in Historical Perspective".
23 See, for example, Markusson et al., *The Social Dynamics of Carbon Capture and Storage.*
24 For an excellent overview of energy storage issues, see Ausfelder et al., "Energiespeicherung als Element einer sicheren Energieversorgung".
25 Yergin, *The Prize,* Chapters 28–30.
26 Hughes, "The Electrification of America: The System Builders".
27 Hughes, *Networks of Power,* 14–17.
28 Cf. Thue, "Electricity Rules", 15–23.
29 See, in particular, Kaijser et al., *Changing Direction.*
30 Högselius, *Red Gas,* Chapter 1.
31 BP, *BP Statistical Review of World Energy 2017;* WNA, "World Uranium Mining Production".
32 Kaijser and Högselius, "Under the Damocles Sword".

33 Rüdiger, *From Import Dependency to Self-Sufficiency in Denmark*.
34 E.g. Mathews, "Biofuels".
35 Högselius, *Red Gas*; Högselius et al., *Europe's Infrastructure Transition*, Chapter 2.
36 Ibid.

Further reading

Hecht, Gabrielle. *Being Nuclear: Africans and the Global Uranium Trade*. Cambridge, MA: MIT Press, 2012.
Högselius, Per. *Red Gas: Russia and the Origins of European Energy Dependence*. Basingstoke and New York: Palgrave Macmillan, 2013.
Hughes, Thomas P. *Networks of Power: Electrification in the Western World 1880–1930*. Baltimore, MD: Johns Hopkins University Press, 1983.

Bibliography

Ausfelder, Florian, Christian Beilmann, Martin Bertau, Sigmar Brauninger et al. "Energiespeicherung als Element einer sicheren Energieversorgung". *Chemie Ingenieur Technik* 87, 1–2 (2015): 17–89.
BP. *BP Statistical Review of World Energy 2017*. London: BP, 2017.
Cordovil, Bruno. "De-electrifying the History of Street Lighting: Energies in Use in Town and Country (Portugal, 1780s–1930s)". In *The Culture of Energy*, edited by Mogens Rüdiger, 30–81. Cambridge: Cambridge Scholars Publishing, 2008.
De Vries, Jan. *The European Economy in an Age of Crisis, 1600–1750*. Cambridge: Cambridge University Press, 1976.
Del Curto, Davide, and Angelo Landi. "Gas-Light in Italy between 1700s & 1800s: A History of Lighting". In *The Culture of Energy*, edited by Mogens Rüdiger, 2–29. Cambridge: Cambridge Scholars Publishing, 2008.
Högselius, Per. "Spent Nuclear Fuel Policies in Historical Perspective: An International Comparison". *Energy Policy* 37 (2009): 254–263.
Högselius, Per. *Red Gas: Russia and the Origins of European Energy Dependence*. Basingstoke and New York: Palgrave Macmillan, 2013.
Högselius, Per, Anique Hommels, Arne Kaijser and Erik van der Vleuten, eds. *The Making of Europe's Critical Infrastructure: Common Connections and Shared Vulnerabilities*. Basingstoke and New York: Palgrave Macmillan, 2013.
Högselius, Per, Arne Kaijser and Erik van der Vleuten. *Europe's Infrastructure Transition: Economy, War, Nature*. Basingstoke and New York: Palgrave Macmillan, 2016.
Hughes, Thomas P. "The Electrification of America: The System Builders". *Technology & Culture* 20, 1 (1979): 124–161.
Hughes, Thomas P. *Networks of Power: Electrification in the Western World 1880–1930*. Baltimore, MD: Johns Hopkins University Press, 1983.
Hughes, Thomas P. *Rescuing Prometheus*. New York: Pantheon Books, 1998.
Hultgren, Åke, and Gunnar Olsson. "Uranium Recovery in Sweden: History and Perspective". SKB Arbetsrapport 93-42, August 1993.
Johnson, Corey, and Tim Boersma. "Energy (In)security in Poland: The Case of Shale Gas". *Energy Policy* 53 (2013): 389–399.

Kaijser, Arne, and Per Högselius. "Under the Damocles Sword: Managing Swedish Energy Dependence in the 20th Century". *Energy Policy* (forthcoming).

Kaijser, Arne, Arne Mogren and Peter Steen. *Changing Direction: Energy Policy and New Technology*. Stockholm: National Energy Administration, 1991.

Kander, Astrid, Paolo Malanima and Paul Warde. *Power to the People: Energy in Europe over the Last Five Centuries*. Princeton, NJ: Princeton University Press, 2013.

Markusson, Nils, Simon Shackley and Benjamin Evar, eds. *The Social Dynamics of Carbon Capture and Storage*. Abingdon: Routledge/Earthscan, 2012.

Klare, Michael T. *Rising Powers, Shrinking Planet: The New Geopolitics of Energy*. New York: Metropolitan Books, 2008.

Kunz, Andreas, and John Armstrong, eds. *Inland Navigation and Economic Development in Nineteenth-Century Europe*. Mainz: Verlag Philipp von Zabern, 1995.

Mathews, John. "Biofuels: What a Biopact between North and South Could Achieve". *Energy Policy* 35 (2007): 3550–3570.

Quader, A.K.M. Abdul. "Natural Gas and the Fertilizer Industry". *Energy for Sustainable Development* 7, 2 (2003): 40–48.

Radkau, Joachim. *Wood: A History*. Cambridge: Polity Press, 2012.

Rüdiger, Mogens. *From Import Dependency to Self-Sufficiency in Denmark, 1945–2000*. Energy Policy (forthcoming).

Smil, Vaclav. *Energy at the Crossroads: Global Perspectives and Uncertainties*. Cambridge, MA: MIT Press, 2003.

Tchalakov, Ivan, Tihomir Mitev and Ivaylo Hristov. "Bulgarian Power Relations: The Making of a Balkan Power Hub". In *The Making of Europe's Critical Infrastructures: Common Connections and Shared Vulnerabilities*, edited by Per Högselius, Anique Hommels, Arne Kaijser and Erik van der Vleuten, 131–156. Basingstoke and New York: Palgrave Macmillan, 2013.

Thue, Lars. "Electricity Rules: The Formation and Development of the Nordic Electricity Regimes". In *Nordic Energy Systems: Historical Perspectives and Current Issues*, edited by Arne Kaijser and Marika Hedin, 213–238. Canton, MA: Science History Publications, 1995.

Thue, Lars. "Connections, Criticality, and Complexity: Norwegian Electricity in Its European Context". In *The Making of Europe's Critical Infrastructure: Common Connections and Shared Vulnerabilities*, edited by Per Högselius, Anique Hommels, Arne Kaijser and Erik van der Vleuten, 213–238. Basingstoke and New York: Palgrave Macmillan, 2013.

Tomory, Leslie. "Building the First Gas Network, 1812–1820". *Technology & Culture* 52 (2011): 75–102.

Vikström, Hanna, Sven Davidsson and Mikael Höök. "Lithium Availability and Future Production Outlooks". *Applied Energy* 10 (2013): 252–266.

Wasp, Edward J. "Progress with Coal Slurry Pipelines (Comparison with Unit Trains)". *Proceedings of the 8th Annual Front of Power Technology Conference*, Oklahoma State University, Stillwater, 1–2 October, 1975.

WNA. "World Uranium Mining Production". www.world-nuclear.org/information-library/nuclear-fuel-cycle/mining-of-uranium/world-uranium-mining-production.aspx (accessed 29 May 2018).

Yergin, Daniel. *The Prize: The Epic Quest for Oil, Money and Power*. London: Simon and Schuster, 1991.

3   Who controls world energy?

Who controls the world's energy supply? Who has the power to influence international energy systems and the movements of energy from one part of the globe to another? These are questions of fundamental importance in the study of energy and geopolitics, and they will be central throughout this book. This chapter starts to explore the theme of power and power relations by taking a closer look at the variety of *actors* who, over the years, have acquired roles of one kind or another in the global energy domain.

By actors, I here mean *individuals* and, more often, *organizations* (businesses, state agencies, non-governmental organizations, etc.). Both undoubtedly have agency in energy and geopolitics. *Countries*, by contrast, do not have agency – and hence should not be treated as actors. While it is common, in public discourse, to talk about "Russia" using energy as political weapon, about "China" as a rising world power and about "Germany" as a leader in the transition to renewables, such simplifications do not bring us far if we want to unravel the nature of power and control in world energy. Rather than studying "Russia", we need to analyse Vladimir Putin's TV appearances, Foreign Minister Sergey Lavrov's manoeuvrings in the international diplomatic arena and Gazprom's export strategies; rather than "China", we need to study the relationship between President Xi Jinping and Sinopec's ex-president Wang Tianpu (who is now in jail); rather than "Germany", we need to study the power struggle between coal importers and radical environmental organizations. We must never treat countries as actors in their own right.[1]

We will look at several categories of actors. We begin by scrutinizing the role of private companies and entrepreneurs, for whom money and profits are usually the prize luring them to engage in energy-related activities. We then turn to the role of state actors, whose motivations are of a more mixed nature, and to state-owned energy companies. We continue by exploring the diverse actors at the regional and local levels, which, perhaps paradoxically, have often been crucial agents in shaping energy internationally. We then discuss the role of environmental organizations, of trade unions and of scientists and engineers. Finally, we turn to the roles of international organizations and the media in energy and geopolitics. At the end of the chapter, we return to the question posed at the outset: who actually controls world energy?

World energy and private enterprise

The dream of striking it rich has clearly been a tremendous motivator when it comes to human engagements with world energy. Private entrepreneurs have made fabulous fortunes from fuel extraction and energy system-building over the past two centuries. Oil is the most obvious example. In the nineteenth century, John D. Rockefeller and Emanuel Nobel, the leading oilmen in the United States and Imperial Russia, respectively, already earned so much money from their business empires that they became the richest of all people in their home countries. Today, energy companies feature prominently among the world's most powerful corporations. As of 2017, both Royal Dutch Shell and Exxon Mobil made it to the Top 10 in *Fortune's* Global 500 list, with several others, notably BP, Chevron and ConocoPhillips, not far behind. Depressed oil prices notwithstanding, Shell had a whopping $240 billion in revenues in 2017, roughly equivalent to the gross domestic product (GDP) of a country such as Finland or Portugal. With such extravagant cash flows comes a great deal of power, and the large oil companies have undoubtedly exercised substantial control over the world's energy supply.

But the "supermajors" – as the largest of the oil companies are called – are just the tip of the iceberg. Private enterprise flourishes in all segments of the global energy industry, and at many different levels. While the large extractive companies are usually the ones we get to hear about in the business news, they are by no means the only ones. Some of the suppliers and sub-suppliers to the large energy producers and system operators are often nearly as powerful as their customers. Schlumberger, for example, a huge company that produces no oil, but offers key services to the oil industry, had $28 billion in revenues in 2016. ABB, a supplier to the electricity industry, earned $34 billion, while Vestas, a leader in wind turbine production, made $8 billion.[2] Nuclear power plant builders, solar panel manufacturers and R&D companies, steel companies producing pipes for the oil and gas industry and steam-turbine makers for thermal power plants are other prominent groups in this category, all of them actors that have been able to shape international energy systems in decisive ways. At times they have found themselves in a central position in energy geopolitics. During the Cold War, for example, US and European exports of steel pipes and gas compressor stations to the Soviet Union, for use in gas system-building there, were at the centre of the debate about Western Europe's energy relations with the Soviet Union.[3] More recently, the political relations between China and the European Union (EU) reached a low-water mark following EU accusations that the Chinese were dumping cheap solar panels on European markets, while equipment supplies for nuclear energy have been at the heart of European and American sanctions against Iran.

And then there are, of course, the many thousands of small- and medium-sized companies that in one way or the other cater to the energy industry for maintenance works, for sub-supplies of various kinds, for consultancy services and so on. As these smaller actors usually operate in the shadow of the world media, we rarely get to hear about them. Yet they are indispensable for sustaining global energy flows. Failure to keep up with maintenance, for instance, is a much more common cause of supply disruptions in international energy systems than the much-publicized political uses of energy as a "weapon" would suggest.[4]

The balance of power between private and public actors in global energy has changed over time. It has also varied strongly from one country or region to another.[5] Up to World War I, world energy was essentially championed by private enterprise, especially in oil and coal; local gas and electricity works, by contrast, were in some cases publicly controlled from the outset. The period from the 1920s to the 1970s became an era of reduced private ownership of major energy companies, for several reasons. One was that the state tended to take a greater interest in fuel resources as "strategic" commodities after World War I and thereby started to view energy as something too important to be left entirely to private businesses. Another was the abolition of capitalist ownership altogether in large parts of the world, from the Soviet Union and Eastern Europe to China and Vietnam. A third reason was the nationalization of foreign-owned companies, especially in the oil industry, in former colonies and other developing countries. The trend was reversed starting in the 1980s, as a wave of liberalization and privatization swept across the globe, further reinforced in the 1990s by the collapse of communism in Russia and Eastern Europe. As a result, many national energy companies were partly or fully privatized and listed on the stock exchange – unless, in the course of the privatization process, they ended up in the hands of one or the other energy "oligarch". Even electricity companies were listed, the most notable case being China's huge State Grid – which is now in Fortune 500's top 10! Saudi Aramco, the world's largest oil company, is also being prepared for partial privatization at the time of writing. At the same time, however, there are signs that the pendulum is now swinging back again, as evidenced by the recent strengthening of state control over energy in Russia, South America and elsewhere.

BOX 3.1 THE RISE – AND FALL? – OF THE "SUPERMAJORS"

In global oil parlance, it is common to talk about the "seven supermajors", comprising ExxonMobil, Chevron, ConocoPhillips, BP, Shell, Total and ENI. Together, these corporations control much of the global flows of crude oil and refined oil products. The supermajors, with the exception of Total and ENI, have a history that goes back to the early days of the oil industry, having been founded in the nineteenth century. In particular, the legacy of John D. Rockefeller, by far the most important oil tycoon of that era, is still very visible in today's power over oil. In 1870 Rockefeller founded Standard Oil, which grew so powerful in the course of the following decades that the US Supreme Court in 1911 enforced its split-up into seven separate companies, with headquarters in different US states. Standard Oil of New Jersey (often referred to as Esso), by far the largest of the "baby-Standards", subsequently evolved to become Exxon. Standard Oil of New York (Socony) became Mobil, and eventually (re)merged with Exxon in 1999. Standard Oil of California (Socal) became Chevron. Chevron also acquired Standard Oil of Kentucky. Standard Oil of Indiana was later renamed Amoco and was eventually integrated into BP, which also acquired Standard Oil of Ohio (Sohio, the original Standard Oil corporate entity that had once led Rockefeller's controversial endeavour). Among the seven supermajors, only the continental European companies of Shell (headquartered in the Netherlands, though with a history of tight connections with Britain), Total (France) and ENI (Italy) do not incorporate any significant remnants of Rockefeller's vast business empire.

What makes the world perceive of the supermajors as powerful actors, apart from their sheer size, is their vertically integrated structure: they are involved in both "upstream" (oil production) and "downstream" (distribution) activities – and everything in between. Most oil companies are large-scale producers of natural gas, too, and some are major players in the petrochemical industry. At the same time, they are impressively globalized in terms of the geographical distribution of their activities. Shell, for example, may have its headquarters in The Hague, and Exxon at Irving, Texas, but the oil systems they (seek to) control are international. As of 2016, Shell was involved in oil and gas exploration and production activities in 48 countries worldwide (of which 11 are in Europe, 19 in Asia and Oceania, 9 in Africa and 9 in the Americas), refinery operations in 12 countries, and chemical plants in 8 countries.[6] Even a comparatively small (non-supermajor) oil company like Equinor (with headquarters in Norway) has exploration and production activities going on in 20 countries worldwide, although all of the refineries it operates are in Norway.[7]

Especially in the developing world, the immense strength of an international oil company, in terms of the capital, expertise and political connections it brings, is usually perceived as overwhelming, and cooperation with such actors often becomes strongly asymmetrical. Such power asymmetries have historically been at the root of many political and commercial controversies over oil. Colonialism is a central term here, as the aggressive expansion of the majors into new potentially oil-rich lands has commonly been interpreted as an extension of Western empire-building ambitions. It is impossible to understand the evolution of the global oil system without taking into account anti-Western and anti-colonialist movements and sentiments in countries such as Mexico, Venezuela, Iran, Iraq, Egypt and Libya. Especially from the 1950s, when global oil supply started growing in an unprecedented way, state actors in the non-Western producing countries began to demand greater control over the oil resources on their territories, a greater say in decision-making over oil and, above all, a larger share of the profits earned by the majors. This ended in the nationalization of oil resources previously controlled by the multinationals. In some cases – notably Venezuela, Iran, Libya and Algeria – this was a radical, revolutionary process. In other cases – notably Saudi Arabia and Kuwait – nationalization and expropriation took the form of more peaceful negotiations. The Saudi Arabian case exemplifies the less radical path to indigenous control of oil: before 1973 the Saudi state had no stake at all in the producing company, the Arabian American Oil Company (Aramco). In several steps, it then gradually increased its ownership to 25%, to 60% and eventually, in 1980, to 100%. In 1988, finally, the Saudi state also took over management and operational control.

All in all, the main historical trend has been one of *reduced* asymmetries between (non-Western) producing countries and the supermajors. The near-total dominance of the latter in international oil has gradually given way to a more balanced relationship, and in many cases even to a marginalization of the foreigners. The 1973/1974 oil crisis marked the culmination of this shift in power. So even if we may get the impression that today's supermajors are aggressive, even ruthless exploiters of oil in a range of countries, their current power is only a pale remnant of the immense force that the majors were earlier able to wield in the developing world. In the twenty-first century, global oil production, and oil imports, are largely in the hands of huge state-owned oil companies. This is the case in North Africa and the Middle East, but also in China, India, Brazil, Malaysia and elsewhere. Moreover, the non-Western state-owned companies have started to challenge the hegemony of the supermajors in non-extractive system activities, too, through investments in refining, transportation and worldwide sales. Saudi Aramco, for example, now has refining operations not only in Saudi Arabia itself but also in the Philippines, China, Japan, South Korea and even the United States. It also employs its own oil tankers for worldwide shipments.

State actors: who are they, and what do they want?

The power of private enterprise in world energy may be immense, but it is rivalled by that of public actors. The latter comprise a colourful palette of people and organizations, representing a multitude of diverse interests and control ambitions. Presidents and prime ministers come to the fore as the most powerful actors in the public realm. At the time of writing, Vladimir Putin in Russia and Donald Trump in the United States are among the most clear-cut examples of government heads whose statements, initiatives, decisions and decrees directly influence international energy relations. They are preceded historically by an intriguing cast of characters, from Mexican President Lázaro Cárdenas and Iranian Prime Minister Mohammad Mosaddegh, who upset the world by nationalizing the oil industries in their respective countries (in 1938 and 1951, respectively), to US President Dwight Eisenhower, who launched the Atoms for Peace initiative in 1953 and thereby transformed the geopolitics of nuclear energy, and Anwar Sadat in Egypt, who set the 1973 energy crisis in motion by launching the Yom Kippur War against Israel.

Presidents and prime ministers are, of course, supposed to represent the interests of the nations they govern. But what exactly is in the interest of the nation? It is possible to discern at least three categories of interests that historically have shaped government decision-making in energy: *foreign policy interests*, *economic interests* and *environmental interests*. A government leader who seriously tries to take into account all three can be expected to try and build international energy relations that support peaceful and constructive relations with other countries, economic development domestically and a healthy (domestic) environment. In actual practice, this is usually easier said than done. It is only under the most favourable circumstances that a head of government finds that working for one of the three interests supports her work for the others. More commonly, she will find that the three contradict each other such that, for example, pursuing economic interests forces her to sacrifice foreign policy interests.

This (potential) conflict of interests becomes more pronounced when we analyse the activities of specific governmental ministries or departments. Such an analysis quickly makes clear that national governments are highly heterogeneous organizational entities. For example, the *foreign office* – in many countries also referred to as the *ministry of foreign affairs* or, as in the United States, the *Department of State* – is a world of its own, featuring a particular organizational culture not shared with other ministries and departments. Foreign offices are as a rule inhabited by ambitious individuals who have received their training at prestigious diplomatic schools, and who see themselves as the ones ultimately in charge of their nation's independence, its respectability in the world and its peaceful relations with other countries. For the foreign office's staff – and here it has a lot in common with the ministry of defence – there is nothing more important than the nation's security, its territorial integrity and its overall reputation in world politics.[8] These are the *core interests* of any foreign office, and they have shaped energy-related initiatives of foreign-policy makers for over a century. Willy Brandt, for example, the West German foreign minister from 1966 to 1969 (after which he advanced to become Chancellor), supported the idea of constructing natural gas pipelines from the Soviet Union to West Germany because he saw it as one in a series of steps that might enable Germany's future reunification.[9] In a similar way, the foreign policymakers in Anwar Sadat's Egyptian government, in the wake of Egypt's military assault on Israel in October 1973, worked hard to bring together the Arab oil exporters in an embargo against the United States and other Western nations because they saw this as a means to win the war and restore Egypt and its prestige in the international arena.[10]

At a nation's *ministry (or ministries) of economy, industry and trade,*[11] energy is viewed from a totally different angle. Like foreign offices, economic ministries take great interest in international energy relations, but the nature of that interest is different. When they take action, they do so to promote fundamental economic variables like GDP, overall standards of living, levels of employment and so on. Those are the core interests of any ministry of economy. Nothing is more important. While its staff does recognize the significance of national security and other typical foreign policy interests, it does not agree that these should be placed "above" issues such as economic development, macro-economic stability, foreign trade balances and the like.

In energy-exporting countries, this is likely to make ministries of economy unwilling to support export embargoes or other uses of the "energy weapon" advocated by a president or a foreign minister, because such measures are unlikely to bring any economic benefits. In energy-importing countries, the ministry of economy's staff will view energy as a key input to domestic economic activities, and they are keen to see domestic industries access sufficient amounts of cheap energy – whether it comes from domestic sources or from abroad. Hence, they may, for example, favour energy imports if these offer a cheaper way to meet national energy needs than domestic extraction. The foreign office is likely to disagree on this, perhaps proposing import duties or bans on imports from certain particularly "unreliable" nations. (In fact, the ministry of economy may also propose import duties, though not so much for national security reasons but in order to stimulate the development of the domestic energy industry – the argument being that such measures will benefit the national economy as a whole.) Employment interests may also be at focus here, as in the powerful European tradition of protecting domestic coal industries from foreign competition.

Foreign policy and economic interests often overlap, so that the government as a whole pulls in the same direction, though with different underlying motivations. In international system-building, new projects are unlikely to materialize unless this is the case; there must be "something in it" from both a foreign policy and an economic point of view. Foreign policymakers may also view economic policy as a potential "lever" in seeking to attain their core goals – and vice versa.[12] But foreign policy and economic interests may also collide, as in the case of hotly debated export embargoes and sanctions of various kinds. Of course, there may also be internal conflicts *within* a foreign office or a ministry of economy. The foreign office may comprise diplomats and advisors representing radically different international relations schools, with some being more "hawkish" and militant and others more conciliatory and cooperative in spirit. The most typical conflict at a ministry of economy, by contrast, is perhaps the one between politicians and advisors who first and foremost have the well-being of domestic *energy producers* in mind (which wish to keep energy prices high) and those closer to *energy users* (which wish to keep prices low). Mitigating between such conflicting interests has historically caused much headache for many a minister or government head.

Finally, *environmental interests* have grown increasingly important in governmental affairs over the past half century – in a way that has strongly influenced the geopolitics of energy. Here, issues such as urban smog, acid rain, global carbon emissions, polluted seaways and loss of biodiversity are at focus. Environmental policymakers view these as more important to take into account than foreign-policy and economic issues. After all, they inform us, national security and economic development are concerns that become meaningless if the world heads towards the environmental apocalypse.

Historically, there has been a huge cultural divide between environmental ministries (and state environmental agencies), on the one hand, and foreign policy and economic policy actors, on the other.[13] The latter have found it hard to acknowledge environmental arguments as legitimate inputs in debates over international energy relations, arguing, for example, that a nation cannot risk compromising its territorial integrity for the sake of combatting local pollution, or that the state should not allow environmental issues to delimit a country's economic growth potential. Conversely, environmental policymakers have typically been leery of foreign policy actors, whom they suspect of serving the dark interests of military-industrial complexes, and of economic policy actors, who are often accused of being allied with brutal capitalist forces for whom pecuniary riches are all that matter. Such debates have loomed large in controversies over, for example, risky nuclear energy programmes as a pathway to oil independence, or investments in polluting lignite mines as a way to counter escalating energy imports. Over the past two decades or so, however, at least a partial rapprochement has occurred in many countries. Both environmental and foreign policy actors nowadays view the transition from fossil fuels to renewables as beneficial. Ministries of economy are increasingly supportive of the transition as well, having concluded that the emerging age of renewable – and "smart" – energy is linked to major economic opportunities.

The main responsibility for energy policymaking in governmental affairs has historically often been with ministries of economy, industry and trade. As the perceived "strategic" importance of energy has grown, foreign offices have usually established a subgroup to deal with energy in the foreign policy context, but the ministry of economy has generally continued to wield the main expertise in energy affairs. In some cases, energy's looming strategic significance has led governments to establish stand-alone energy ministries; the US Department of Energy is the most well-known. Countries where energy dominates the national economy also feature strong energy ministries. The most powerful oil exporters even have specialized petroleum ministries. At the same time, there has been a trend in recent decades towards moving the main responsibility for energy policymaking to ministries of climate and environment; this is the case in many smaller countries with high-profile environmental agendas, like Sweden, Denmark and Costa Rica. These organizational patterns hint at the interests that are prioritized by different governments.

In pseudo-democratic and authoritarian states – or, more generally, in states featuring a strong, unchallenged political leader – one might get the impression that there are no controversies within the government over its energy-related interests. Nothing could be more wrong. Putin's government in Russia, for example, is torn between its economic interest in earning money from oil and gas exports to its ex-communist neighbours, and its foreign policy interest in using energy to punish the same countries in cases where these criticize Russian political behaviour or orient themselves too much towards the West. In Xi Jinping's China, imports of natural gas are viewed as potentially dangerous from a foreign policy point of view, but as necessary when it comes to coping with the extremely severe air pollution that plagues almost every major Chinese city. The only difference, really, between democratic and authoritarian countries in terms of conflicting state interests is that we may read about them in the news in the former case but not in the latter.

The double identity of state-owned energy companies

A special category of state actors, of immense importance in energy and geopolitics, is that of state-owned energy companies. These have one leg in the state apparatus and one in the world of business. The relative strength of the two legs varies a lot from case to case. At one extreme, we here find energy companies that are officially classified as governmental ministries and which thus formally have nothing to do with the world of business. This was the case in Maoist China and the former Soviet Union, whose energy industries were fully in the hands of their respective state administrations. Up to the late 1970s, Chinese coal production was thus in the hands of the Ministry of Coal Industry, while oil production was the responsibility of the Ministry of Petroleum Industry.[14] The Soviet government comprised a

Ministry of Electrification, a Ministry of Oil Industry, a Ministry of Gas Industry and so on. The Ministry of Gas Industry (Mingazprom), to take one example, was not merely a state bureaucracy set to *regulate* or develop *policies* for the gas industry; rather, the ministry was itself in charge of producing and transporting gas, of building pipelines, of searching for new gas deposits and so on – it was a system-building agency as well as an operator of the entire gas system. It was not formally a "company". However, it often behaved very much like one, seeking to promote gas production, expand the system, improve efficiency – and sell as much gas as possible. After the collapse of the Soviet Union, the ministry was transformed into the powerful gas company Gazprom.[15]

At the other end of the spectrum, we find organizational entities that are no different from a "normal" joint-stock company – save the fact that the state holds a majority of the shares. Norway's Equinor (formerly Statoil), for example, is controlled by the Norwegian government, which owns 67% of the shares, but the remaining 33% is owned by private capital and the company is listed on the Oslo stock exchange. Similarly, 71% of the Sinopec Group – modern China's largest enterprise and the world's fifth largest company in terms of revenues – is currently owned by the Chinese state, while the remaining 29% is privately owned; the company is listed on several of the world's leading stock exchanges. But there are also state-owned companies that are wholly controlled by governments. Sweden's Vattenfall, for example, one of Europe's leading electricity companies, is owned 100% by the Swedish state, while the National Iranian Oil Company (NIOC) is 100% owned by the Iranian government.

BOX 3.2 THE VOLATILE CAREERS OF THREE STATE-OWNED ENERGY COMPANIES

BP is currently one of the world's largest oil companies and one of the seven supermajors. Its history goes back to the oil concession that William Knox D'Arcy, an English businessman, acquired in Persia in 1901. D'Arcy initially failed to find any oil there. Having lost a fortune in exploring Persia, he eventually sold most of his rights to the Scotland-based Burmah Oil Company. Burmah's support made it possible to continue exploration, eventually resulting in the discovery of an immense oil field in 1908. Shortly after that discovery, Burmah set up a Persian subsidiary, named the Anglo-Persian Oil Company, to manage the field's exploitation. The first oil was delivered in 1912. At that time, oil's strategic importance was becoming increasingly evident, and in this context Anglo-Persian soon attracted the attention of the British government. Winston Churchill, serving as First Lord of the Admiralty, took the fateful decision to transform the British Navy by abandoning its reliance on coal and steam and, instead, building up a fleet of oil-propelled warships. This gave rise to supply concerns, since Britain had no domestic oil deposits. Seeking to avoid becoming dependent on private enterprise and, in particular, American-controlled oil companies, Churchill approached Anglo-Persian. By the outbreak of World War I, he had managed to acquire a controlling stake in the company. Through that stake the British government also controlled 47.5% of the important Turkish Petroleum Company, which in 1927 found large oil deposits at Kirkuk in northern Iraq.

In 1935, Anglo-Persian was renamed the Anglo-Iranian Oil Company. Britain's control of the company gave rise to controversies and fuelled Iranian nationalism. The conflict escalated after World War II, and especially after 1950, when the British government refused to accept the 50-50 profit-sharing scheme that by now was becoming a standard formula in Middle Eastern oil operations. In March 1951, newly elected Iranian Prime Minister Mohammed Mosaddegh took the decision to nationalize the country's oil industry. The assets in Iran belonging to Anglo-Iranian formed the basis for a newly formed National Iranian Oil Company (NIOC), and the British were thrown out. Following a coup two years later, orchestrated by the West, Mosaddegh was removed and the British were allowed to return. They resumed work in Iran's oil fields and refineries, although they no longer controlled the country's oil industry as before. Anglo-Iranian, still controlled by the British government, was now transformed into the British Petroleum Company, which set out to broaden its activities in geographical terms so as to avoid its dependence on Iranian oil. One of its major feats became the 1968 discovery of the huge Prudhoe Bay oil field in Alaska. It also became very active in the North Sea, as well as in Latin America, Africa and Southeast Asia. However, governmental control of the company was gradually reduced. As of 1970, the government still held a 48% stake in the company. Under the Labour government in office from 1974 to 1979, the share fell to 26%. Then, under Margaret Thatcher's conservative government, all remaining shares were sold to the private sector in 1987.[16]

Gazprom is Russia's most well-known – and controversial – state-owned energy company. It was originally founded in 1956 as a spin-off from the Soviet Union's Oil Ministry. Initially it was known as the Main Directorate of the Gas Industry, but as gas production was scaled up it was upgraded to a ministry in 1965. The head of the directorate, Alexei Kortunov, thus became Minister of the Gas Industry. During the next 25 years, the ministry's – Mingazprom's – activities grew at a remarkable pace; by the 1980s, it had become one of the most powerful ministries in the Soviet Union, challenging the country's huge oil and coal ministries. In 1989, following attempts to reform the Soviet economy, Gorbachev's government decided to corporatize the ministry, turning it into the State Gas Concern Gazprom. Viktor Chernomyrdin, who had been Minister of the Gas Industry since 1985, emerged as the head of this state enterprise.

Through the dissolution of the Soviet Union in late 1991, the concern faced an existential crisis as it was split up into several smaller enterprises, each one with a national coverage, such as Turkmengazprom (later renamed Türkmengaz) and Ukrgazprom (now Naftogaz). The majority of its assets and value, however, remained in what was now the Russian Federation. Gazprom also took over the profitable gas export business, which in Soviet times had been controlled by the Ministry of Foreign Trade. Through the geopolitical changes, this business came to include large-scale gas exports not only to Western and Central Europe but also to what the Kremlin referred to as its "near abroad" – the non-Russian ex-Soviet republics, of which Ukraine was by far the largest. The close links between gas and politics in these chaotic times were epitomized by President Boris Yeltsin's appointment, in December 1992, of Gazprom chief Chernomyrdin as Russian Prime Minister. Shortly afterwards, however, Gazprom was turned into a joint stock company, and a majority of the shares were sold to Russian citizens (through a voucher privatization scheme), to Gazprom employees and, on a smaller scale, to foreign investors. The company was now also listed on the stock exchange. The state retained 38% of the shares – a sizeable, but not a controlling stake. After becoming Russian President in 2000, Vladimir Putin set out to change this and, through skilful financial engineering and different forms of pressure on the non-state shareholders, managed to raise the state's interest to over 50%. Hence, the Russian government once again controls Gazprom.[17]

Vattenfall was founded in 1909 as Sweden's state electricity board. For a long time, it enjoyed the status of a state agency rather than a state-owned company. Vattenfall became the main actor in harnessing northern Sweden's immense hydropower potential. Its aggressive expansion there gave rise to numerous conflicts and controversies, partly because dam construction made life difficult – or impossible – for other economic actors in the area, notably those engaged in fishing and log-driving, and partly because the system-builders totally neglected the interests of the indigenous Sami population. However, hydropower from the north, produced by Vattenfall, became a huge success story from a national economy point of view, supplying Swedish energy-intensive industries with massive volumes of cheap electricity. In the process, Swedish equipment manufacturers such as ASEA (now ABB) emerged as world leaders in fields like high-voltage transmission technology, which was direly needed for moving electrical energy from Sápmi in the north to the industrialized regions of southern Sweden.

In the interwar era, regional competitors complained about Vattenfall's growing dominance in the electricity market, but World War II served to strengthen national cooperation. After the war, it was agreed that Vattenfall take sole responsibility for the operation of the nation's unified high-voltage grid. The agency also pushed with great enthusiasm for the creation of an interconnected Nordic electricity system. That system became a showcase of Nordic cooperation during the Cold War era. By the late 1980s, however, electrification was more or less completed and Vattenfall's sales stopped growing. The neo-liberal trend in world economy and politics drove discussions about privatizing the agency. The idea was to let the agency diversify beyond the electricity sector or, alternatively, expand to foreign markets. In 1991, the government decided to corporatize Vattenfall by turning it into a joint-stock company. This was widely interpreted as a first step towards privatization. As of today, however, the state continues to be the sole owner. From the late 1990s, under the lead of Lars G. Josefsson, the company embarked on a radical foreign expansion. Josefsson skilfully engineered the acquisition of key electricity companies in Poland and, above all, Germany. There, the Swedish state-owned company gained control over the electricity systems of both Hamburg and Berlin, and of the entire ex-East German transmission grid – along with a set of lignite mines and lignite-fuelled power plants. The German acquisitions became highly controversial among the general public in Sweden. Yet the government took a passive stance, and actual control of the company, 100% state ownership notwithstanding, appeared to be in the hands of Josefsson and his managers. For a number of years, the company made huge profits. Then, in the 2010s, it started to make disastrous losses, mainly due to low electricity prices and unsuccessful further expansion into the Netherlands. The German government's decision to prematurely phase out nuclear energy following the 2011 Fukushima nuclear accident further worsened things, since the company operated several German nuclear power plants. This led to an enflamed conflict with the German government, which is not yet resolved at the time of writing. Vattenfall's future now looks highly uncertain.[18]

What interests do state-owned energy companies have? What do they want? What are their agendas? If we ask their managers, in most cases these will tell us that they are, in fact, no different from a "normal" company: they generate value for their shareholders. In this sense, it would seem that their activities and strategies are shaped mainly by economic interests. But there may be other, competing agendas. For example, the managers of a state-owned enterprise may take a patriotic stance, pointing to the great importance of what their company is doing for their nation's overall security, development and prosperity. This is especially common in cases where the company generates big profits or pays large sums in taxes or royalties to the government, thus strengthening the state budget. The state is, of course, happy to receive such streams of revenue. But governments may also be tempted to mobilize their companies for non-economic purposes. The Russian government's alleged use of Gazprom as an instrument for foreign policymaking is an iconic – and contested – case. The Chinese state's agenda in relation to the activities of its state-owned oil companies in Africa has been interpreted in similar terms. There are many other examples.

More often than not, however, governments may find it difficult to impose any political agenda on the operations of their state-owned energy companies. This is because company managers are typically not inclined to become instruments of foreign policy. They dislike such connections between business and politics – except where these can help them to gain a competitive edge. If governments try to intervene too much, the managers may choose to resign, and the state may find it difficult to recruit competent replacements. In Sweden, for example, there have been many political initiatives aimed at mobilizing Vattenfall to speed up the transition to renewable energy – both at home and abroad. But apart from its historical investments in Swedish hydropower, Vattenfall today remains almost totally dependent on nuclear power and fossil fuels as the basis for its electricity production. The board of directors is able to sustain this profile by pointing to regulations such as the state's request that the company make a decent profit every year – and so far the easiest way to make profits has been to linger in the fossil era. The Swedish public was outraged by Vattenfall's acquisition of several large German lignite mines in 2001, but it was only in 2017 that the government finally managed to force the company to sell off the mines.

The Vattenfall experience, which shows many similarities to corresponding stories about Finnish, French, Italian, Spanish and other European state-owned companies,[19] suggests that there is often a problematic power asymmetry between large state-owned companies and the governments that are supposed to control them. A state-owned company may have hundreds of experts work out their strategies and responses to attempts by the state to intervene in its business. The government, by contrast, usually has only a few people look after the company – perhaps none of whom is an expert on the energy industry.

BOX 3.3 TOWARDS STATE CONTROL OF GLOBAL COAL EXTRACTION

The global coal industry has seen several shifts in the power balance between private and public enterprise. In the nineteenth century, business actors and private capital played a dominant role in coal industries almost everywhere. In contrast to oil, where a few extremely powerful actors came to dominate the world market, coal production was in the hands of a very large number of companies, none of which was particularly dominant. This was the case in both Europe and North America as well as in Asia. At times, notably during the interwar era, groups of coal companies joined forces to create cartels or syndicates of one kind or another – the most famous being the Rhenish-Westphalian Coal Syndicate in Germany – but occasional warnings that a "Coal OPEC" might be in the making never really materialized.

After World War II, coal production was nationalized in Britain and France and, of course, also in Mao's China and the communist countries of Eastern Europe following their incorporation into the Soviet sphere of interest. From the 1980s, the trend turned towards the privatization of state-owned coal companies in both Eastern and Western Europe. In recent decades, however, coal production in Western Europe (where the hard coal industry has essentially ceased to exist), Eastern Europe and North America has been dwarfed by the spectacular rise of the Chinese and Indian coal industries, which now together account for a staggering 54% of total world coal production,[20] and whose coal companies are state-owned. In other words, the global shift in the coal production geography has had the effect – if we take a global perspective – of moving the power over coal from the private to the state sector. To illustrate this, we may compare the revenues of China's largest coal producer, the Shenhua group, which in 2014 amounted to $53 billion, with those of the world's largest privately controlled coal company, US-based Peabody (with headquarters in St. Louis, Missouri), which amounted to a mere $4 billion in 2016.[21]

Regional and local actors

The geopolitics of energy is often thought of as a domain for grown-ups: for large, multinational energy companies and powerful national governments. However, smaller and less powerful actors at the subnational (regional and local) level have often played surprising and unexpected roles in shaping and controlling international energy systems.

Take the case of Bavaria, one of Germany's federal states. Although the federal states of Germany are not supposed to have any foreign policy of their own, in the 1950s and 1960s regional Bavarian actors were almost as active on the international energy arena as the federal government in Bonn. The point of departure was Bavaria's relative economic backwardness in the immediate post-war era, which local politicians thought was linked to the region's lack of coal deposits and its dependence on coal "imports" from northern Germany. Under the lead of its Minister of Economy, Otto Schedl, Bavaria actively pursued an energy strategy that aimed to make Bavaria independent from these "imports". Crude oil from North Africa and the Middle East became a key tool in what amounted to an intense struggle for Bavarian coal independence. In order to access foreign crude without becoming dependent on northern Germany, Schedl forged cooperation with oil exporters in North Africa and the Middle East, and with oil companies in Italy, Austria and Switzerland, through which the fuel would have to be transported. By the late 1960s, oil unloaded in the Mediterranean ports of Genoa and Trieste was being pumped across the Alps to landlocked Bavarian refineries. Later on, Bavaria followed this feat up by championing West German imports of Soviet natural gas, which in a similar vein reached Bavaria without transit through northern Germany.[22]

Regional and local actors often discover that they are able to mobilize their geographical peculiarities for their own purposes in international energy relations. This is especially the case when they happen to be located on a transit route for energy on its way from one part of the world to another. Take Gotland, the largest island in the Baltic Sea. It has historically enjoyed a strategic – but also vulnerable – position in military and international affairs. Currently it is home to no more than 58,000 people. Like many small municipalities in northern Europe, it eagerly looks out for opportunities to strengthen the local economy and fight depopulation and unemployment. Among other things, the municipal administration has been highly interested in large-scale infrastructural projects in its vicinity, whereby especially the construction of the Nord Stream natural gas pipeline, designed for scaling up Russian natural gas exports to continental Europe, early on captured the islanders' imagination. When the second phase of this controversial pipeline project started in 2016, the local politicians struck an ingenious deal with the Nord Stream pipeline company (owned 51% by Gazprom and with German, Dutch and French gas companies as minority shareholders): it agreed to rent out the local deep-sea port at Slite, on the island's east coast, for the purpose of facilitating temporary storage of gas pipes and related equipment during the construction phase. In addition to paying rent, the company would employ local port workers for a variety of tasks. But the initiative quickly ran into trouble when the national Swedish government heard about the deal. The government intervened, objecting to the project on national security grounds. After a lively debate and intense negotiations between the municipal leaders and the national

Swedish foreign and defence ministers, the project was cancelled. However, the national government agreed to pay generous compensation to the municipality.[23]

Other local actors see opportunities not in the *creation* of new transnational energy systems, but in their *destruction*. Oil and gas pipelines, electricity transmission lines, energy storage facilities and other components of international energy systems are vulnerable to sabotage from local terrorist and guerrilla groups, who in this way may cause considerable havoc in cross-border flows of energy. During the 2011 Arab Spring, for example, local Bedouin tribes in the Sinai desert blew up the terminal of a pipeline carrying natural gas from Egypt to Israel, "disrupting the flow and causing millions of dollars of damage". The action was repeated at least 12 times. The violence was interpreted as a desperate way for the Bedouin to show their dislike for the central Egyptian government in Cairo, which for years had treated them in a discriminating way, worsening poverty and alienation.[24]

Across the Atlantic, Native Americans and Canadian First Nations have often been highly critical of – and have taken action to prevent – oil and gas system-building on their territories. One of the most hotly debated cases concerns the expansion of the much-publicized Keystone oil pipeline system, which connects crude oil and tar sand resources in Canada and tight oil production in the northern United States with the oil pipeline hub at Cushing, Oklahoma, as well as with oil refineries in Illinois and Texas. When the US House of Representatives in November 2014 voted in favour of a new extension to the system – dubbed Keystone XL – the Rosebud Sioux Tribe, living in one of the areas through which the pipeline would pass, reacted vehemently. "The House has now signed our death warrants and the death warrants of our children and grandchildren", the president of the tribe, Cyril Scott, stated angrily.

> The Rosebud Sioux Tribe will not allow this pipeline through our lands. We are outraged at the lack of intergovernmental cooperation. We are a sovereign nation, and we are not being treated as such. We will close our reservation borders to Keystone XL. Authorizing Keystone XL is an act of war against our people.[25]

The activism contributed significantly to the Obama administration's final decision not to approve – at least not for the time being – the pipeline's construction. That decision made other actors furious – not only the oil and pipeline companies involved but also political actors who thought that halting the project might jeopardize US energy independence ambitions. President Trump, once in power, sought to reinvigorate the project, but at the time of writing it still appears uncertain whether the pipeline will ever materialize, as environmental organizations managed to bring the case to court in early 2018.[26]

Environmental organizations and trade unions

Environmental organizations, as demonstrated by the Keystone XL case, constitute an additional group of influential actors. The geopolitical impact of environmental activism can be considerable, especially when it comes to preventing (parts of) new energy systems from materializing or forcing system-builders to take into account the environmental dimension. For instance, environmental NGOs have played key roles in preventing nuclear power plants, hydroelectric stations and oil and gas pipelines from being built. They have also helped to enforce the shutdown of existing coal mines, nuclear facilities and so on.

Some of the most dramatic cases of environmentalism in relation to energy geopolitics play out in what Jane Dawson has called "eco-nationalism". The struggle for national independence among several of the former Soviet Union republics is particularly illustrative here. In the post-Chernobyl years, Moscow's plans to further expand the Ignalina nuclear power plant in the Lithuanian SSR, for example, spurred the rise of an anti-nuclear movement. As the movement evolved, however, it increasingly came to focus less and on nuclear energy and more on Lithuanian autonomy from Moscow in overall political decision-making. In the end, the anti-nuclear protests evolved into a national independence movement. Once independence from Moscow had been achieved in 1991, the Lithuanians turned much more pro-nuclear. When the EU sought to force the country to close down the existing reactors at Ignalina in connection with the nation's EU accession in 2007, the Lithuanians objected – albeit in vain.[27]

Figure 3.1 Leave Vattenfall's coal in the ground! Greenpeace activists and activists from the citizen initiative "We are the limits" protest in front of the Swedish embassy in Prague on 25 April 2016, opposing the selling of lignite coal mines and associated power plants in Germany to the Czech company EPH. The activists encouraged Vattenfall not to sell, but to shut down the lignite mines. Similar protests took place in at least 13 European countries. Photo: Petr Zewlakk Vrabec/Greenpeace.

At the international level, world energy would not be the same without powerful environmental organizations such as Greenpeace, which has offices in 40 countries worldwide and resembles a muscular international organization more than an environmental NGO. Through its highly visible – and controversial – actions, which have involved everything from scaling offshore oil rigs to breaking into nuclear power plants, Greenpeace influences energy both locally and on the level of global discourse. Another notable organization with international reach is the Bellona Foundation, founded in 1986, which started out protesting against the Norwegian oil and gas industry's environmentally dangerous activities, but subsequently expanded both geographically and thematically. Among other things, it has criticized Russia's mesmerizing nuclear energy complex on environmental grounds. In the process, Bellona's key nuclear expert, Alexander Nikitin, a former Soviet naval officer, was arrested by the Russian Federal Security Service (FSB) and charged with treason through espionage.[28]

Trade unions constitute another category of actors that have played critical roles in world energy. This has been the case especially in coal, where miners and other coal workers early on discovered that they were able to exert pressure on their employers – and even on national governments – by organizing themselves and, when necessary, laying down their work. There have been many legendary coal strikes in history, with workers seeking higher wages, lower work hours, better work conditions and greater influence more generally.

Britain stands out here. British coal-miners early on earned a reputation of being militant, with major strikes taking place in 1890, 1893, 1910 and 1912. Since Britain was a powerful coal exporter at that time, the strikes had international repercussions in the form of coal shortages and increased prices in the importing nations.[29] Further British coal strikes occurred during the latter half of the twentieth century, as more and more mines, facing competition from foreign suppliers and, above all, from other energy sources (notably oil), were being shut down. One of the worst strikes occurred in parallel with the 1973/1974 oil crisis, at which time Britain faced an energy emergency due to the combination of the Arab oil embargo and the striking coal workers.[30] The coal strike of 1984/1985, led by the National Union of Mineworkers (NUM) against the National Coal Board (NCB) and Margaret Thatcher's conservative government, also became a deeply traumatic event.

The most notable American union-led coal strike occurred in 1902 in eastern Pennsylvania's anthracite coal fields. Organized by the United Mine Workers of America, the strike threatened to jeopardize the energy security of major East Coast cities that depended on this fuel. Over the years, coal strikes have, in a similar vein, threatened to disrupt supplies from mines in countries such as South Africa, Spain and Australia.

Port workers have also gone on organized strikes, disrupting energy exports and imports. In Sweden, for example, dock-workers went on strike throughout the country in April 1908. The conflict escalated when some stevedoring companies brought in strike-breakers from Britain. In Malmö a boat on which strike-breakers were housed was hit by a bomb, killing one strike-breaker and severely injuring seven others. The Swedish Employers Federation reacted by giving notice of a general lockout. This forced the government to appoint a conciliation commission, which was able to achieve an agreement between the employers and the trade unions. After more than three months, the coal supply resumed – but the workers had now clearly demonstrated their control over an increasingly vital flow of fuel.[31]

The trade unions' striking habits subsequently spread from coal to other energy industries. In the early twentieth century, oil workers led the way in forming a Russian labour movement. Much later, in the 1980s, Norwegian oil and gas workers in the North Sea went on strike, causing disruptions in gas exports from Norway to continental Europe. The action, which was repeated on several occasions, made natural gas analysts re-evaluate the relative risks involved in importing gas from Norway and from the Soviet Union. Some observers acidly noted that the Soviets were more trustworthy, and that European importers could be sure that the Soviet political leadership, in contrast to the Norwegian government, would not allow any labour strikes to disrupt fuel shipments.[32]

Figure 3.2 Joseph Stalin in Baku, 1908. From the late 1870s, Baku in the southernmost corner of Tsarist Russia spectacularly emerged as one of the world's most important oil production sites. With a steadily growing population of oil industry workers, it also became a hotspot for socialist revolutionary activity. Some of the twentieth century's most influential communist leaders started their political careers in this environment, organizing powerful labour movements at a time when oil was just becoming a strategic resource. This drawing depicts a young Stalin addressing a meeting of Baku oil workers. Source: Mary Evans/ John Massey Stewart Collection.

Scientists and engineers

It is perhaps unfair to introduce scientists and engineers as an actor category only here towards the end of this chapter; their activities have often been decisive in shaping and transforming world energy. As a rule, scientists and engineers represent some private company, state agency or university. Yet there is often good reason to highlight them as actors in their own right. As scientific insights and technological advances have become ever more critical in sustaining and advancing the global energy system, their impact on energy and geopolitics has been vast. Hitler's scientists and engineers, for example, led Nazi Germany's frantic efforts to develop synthetic liquid fuels based on coal – the purpose being to make the country independent of oil imports. The scientists working on the US Manhattan Project, meanwhile, paved the way for the nuclear age. Such controversial developments aside, the gradual, slow advances in scientific fields such as geology, geophysics, petrochemistry and biology have transformed world energy in unprecedented – and often deeply disruptive – ways over the past hundred years. These advances have created tremendous new opportunities but also risks. For example, post-World War II improvements in geological surveying techniques and growing efficiency in oil refining generated a problematic oversupply of oil. As a result, oil became extremely cheap – with well-known consequences in terms of skyrocketing oil dependencies and new geopolitical relations.[33]

Historians of science argue that it is impossible to separate scientific activities from geopolitics. It is not the case that scientists first make a discovery or two, followed by attendant geopolitical impacts; rather, science is geopolitical in and of itself. For example, a geological study of the Arctic region cannot be framed as a "purely scientific" undertaking; in view of the hopes of oil companies and of governments for the Arctic to deliver a greater share of world hydrocarbons in the future, combined with the military-strategic importance of the region, doing science there is now inevitably a geopolitical act. Moreover, state-led investment in higher education and scientific competence-building is not necessarily as innocent as it may appear at first glance; rather, scientific strengths in fields such as the geophysical sciences have historically been intimately linked to the quest for national energy independence and control over domestic prospecting and exploration activities.[34]

In the future, the role of scientists and engineers in shaping global energy developments is bound to increase further, as the world turns to novel energy technologies such as second-generation biofuels, "Generation IV" nuclear reactors, new materials for solar cells, new automotive technologies, new control systems for electricity distribution and so on – all of which are likely to be accompanied by transformed geopolitical relations.

International organizations in energy

International organizations have historically shaped the geopolitics of energy in fascinating ways. We may here distinguish between international organizations that are explicitly dedicated to energy, and general international organizations of relevance for the energy domain. Those dedicated to energy may in turn be divided into bodies that are focused on *a particular energy source,* and others that take a wider interest, targeting *the energy domain as a whole.*

Their importance and relevance have varied over time. In oil, international organizations played a negligible role until 1973, when the Organization of the Petroleum Exporting Countries (OPEC) suddenly came to the fore as the most powerful force of all. OPEC, with headquarters in Vienna, had originally been formed back in 1960, and its ambition had always been to shift the balance of power away from the international oil companies and the rich importing nations in the West towards the producing countries of the "Third World". The source of inspiration came from historical success stories of cartel-like operations. In particular, OPEC took inspiration from the Texas Railroad Commission in the United States, which since the early days of the oil industry had played a central role in regulating much of the US oil industry – and which in the 1930s, when an oil price collapse threatened the industry, had managed to keep oil prices high by coordinating the oil companies' production and shipments. This was precisely what OPEC aimed to do – but on a much greater scale. They failed in the 1960s because production in other parts of the world was so abundant, but succeeded in the 1970s when non-OPEC oil supplies became scarce. In the absence of that scarcity, the 1973 Arab oil embargo would not have had much effect.[35] From the 1980s to the early 2000s, the organization's leverage was reduced as new oil finds were again recorded in many parts of the world and OPEC's share of global oil production declined. More recently, many non-OPEC oil supplies, especially those in the North Sea, again face depletion, causing new fears in the West of growing OPEC power – especially in view of recent attempts by OPEC and Russia to forge an alliance.

While OPEC is linked to the geopolitics of decolonization and North-South relations, the European Coal and Steel Community (ECSC), set up in 1951, was created against the backdrop of World War II and the emerging East-West conflict. Political visionaries thought that coal (and steel) could help promote not only energy security and industrial development but also Western Europe's political cohesion. The focus was initially on coping with coal scarcity, which had been significant in the immediate post-war years – most radically illustrated by the 1947 British "coal famine" – and on keeping up the mood (and wages) of Western Europe's coal workers so that they did not turn into socialist revolutionaries. Subsequently the organization was forced to deal with abundance rather than scarcity, as European coal users found supplies from overseas more attractive than internally produced coal. The ECSC's relevance declined following the massive transition from coal to oil in most European countries. However, it is still celebrated by diplomats and "Eurocrats" as a visionary political project and as the main predecessor of today's EU.[36]

Figure 3.3 IAEA inspectors at work in the Czech Republic, November 2012. The inspectors seal the spent fuel pond, which is situated next to the reactor at Dukovany Nuclear Power Plant. Photo: Petr Pavlicek/IAEA.

In the field of nuclear energy, Western Europe's governments early on took the initiative by setting up the European Atomic Energy Community (Euratom) in 1957. But in the end Euratom was not able to influence international nuclear relations to the extent its visionaries had anticipated. More important, to this day, is the International Atomic Energy Agency (IAEA), which, just like OPEC, has its headquarters in Vienna. In its early days, it had the ambition to play a role in coordinating the international nuclear fuel supply system. While this failed, it is now mainly known as the world's nuclear "watchdog", seeking to make sure that information about nuclear accidents and incidents is properly circulated among the global community of nuclear engineers, and that nuclear materials of potentially military use do not proliferate.[37]

International organizations are less prominent in the natural gas industry. Here, transnational cooperation has been dominated by bilateral agreements, linked to the construction of long-distance pipelines and long-term gas supply contracts. The International Gas Union (IGU) has sometimes acted as a facilitating actor, offering an arena for meetings and general discussions, especially among experts, but occasionally also involving government representatives. In recent years, the EU's role in natural gas has increased, especially after all EU gas transmission companies joined forces in the European Network of Transmission System Operators for Gas (ENTSOG), whose tasks are regulated by an EU directive. Apart from facilitating the creation of a more liquid gas market in the EU, an important motivating force in setting up this new organization is the perceived need to strengthen the EU's gas importers *vis-à-vis* external exporting nations such as Russia and Algeria.

The International Energy Agency (IEA) is the most important international organization in the *general* energy domain. Historically it was primarily linked to oil, being set up in 1974 in direct response to the first oil crisis, but it has subsequently widened its coverage to include other energy sources as well. A sub-organization of the Organization for Economic Cooperation and Development (OECD), it essentially represents the interests of the world's richest countries, most of which are also heavily import-dependent. Hence, the IEA has often been construed as a counterweight to OPEC. In this sense, it has also served to polarize the relations between oil exporters and oil importers. (The French government opposed the IEA's creation, proposing instead a joint Arab-European organization, in which exporters and importers would cooperate; however, the US-led IEA initiative soon prevailed.)

The energy-specific organizations aside, a number of *general* international organizations have also shaped the geopolitics of energy. One of the first was the League of Nations, nowadays mainly known as the predecessor to the United Nations. It was set up after World War I. Taking inspiration from what was perceived as fruitful cooperation among the allied countries during the war, which, among other things, had included coal distribution agreements, the League began discussing possible international cooperation in electricity in the 1920s. It adopted a convention on electric power transmission and another one on hydropower in transnational river basins, which argued that choices made for new transmission lines and for the construction of hydropower plants should be based on technical considerations, and not so much on political ones. The initiative seemed timely, as many countries were just in the process of radically scaling up their electricity supply systems. However, the conventions were ratified by only a handful member states, and thus the League failed to gain any significant influence over electricity's internationalization.[38]

After World War II, the United Nations Economic Commission for Europe (UNECE), set up in 1947, took a far-reaching interest in international energy issues. Like the League of Nations, one of its main ambitions was to prevent national political borders from getting in the way of "technically rational" flows of fuel and electricity. The North Atlantic Treaty Organization (NATO), which, of course, is essentially a military international organization, also became very active in this field. Like UNECE, it promoted energy's internationalization and international cooperation, but only among its member states; it thus worked actively against any system-building ambitions that involved communist countries. On the other side of the Iron Curtain, the Council for Mutual Economic Assistance (CMEA or COMECON) operated according to a similar logic, forging cooperation in energy among the communist countries while excluding the participation of capitalist nations.[39]

Two recent examples of general international organizations relevant for the energy field are the Gulf Cooperation Council (GCC), set up in 1981, and the Shanghai Cooperation Organization (SCO), established in 2001. GCC aims to, among other things, create a unified electric power grid among its member states (currently comprising Saudi Arabia, Bahrain, the United Arab Emirates, Kuwait, Oman and Qatar). SCO originally counted China, Russia, Kazakhstan, Kyrgyzstan, Uzbekistan and Tajikistan as its members, but in 2017 India and Pakistan also joined in. SCO thus covers much of the Eurasian landmass. Joint energy projects are very much at the focus of SCO's work, especially in terms of international oil and gas system-building.[40]

The role of the media in energy and geopolitics

Finally, a few words should be said about the media's role in energy and geopolitics. This would perhaps merit a book in its own right, given the huge number of newspaper articles, TV programmes, Twitter posts and so on that on a daily basis flood us with (real and fake) news about world energy developments – starting with the most recent updates on the Brent oil price and ending with various violent clashes somewhere in the world said to be linked to fuel extraction. It is the media that – for better and for worse – shape the general public's perceptions of what energy and geopolitics are about. Actors of various breeds often skilfully exploit the available news channels as tools in their struggle for control over national and global energy systems. But the media may also put healthy pressure on energy actors, both public and private, by spotlighting morally dubious behaviour, uncovering crime and corruption, reframing discourses and problematizing international energy issues in useful ways.

However, only certain types of narratives and stories are likely to attract a large readership, generating views of energy and geopolitics that are often strongly biased. Only certain stories can be "sold" to the general public, and in many countries news about energy is further filtered through (self-)censorship. For example, we are likely to read much more about the sudden eruption of international conflicts relating to energy than about the slow, "boring" process of conflict resolution. As a result, we get a distorted view that exaggerates conflicts *vis-à-vis* friendly and cooperative relations. In a similar vein, reading about multinational companies in energy, we are more likely to learn about the many scandals these are repeatedly involved in and about their scrupulous behaviour in one or the other part of the world, and not so much about their contributions to making the world a better place. Stories of the latter kind, which clash with the popular image of transnational corporations as brutal and even evil forces in the world arena, do not fit the critical mind-set of Western readers.

EU-Russia relations have long been key to the West European debate about energy and geopolitics, and the media coverage has been substantial. However, in the West, it is now almost impossible to "sell" anything else to the general public than stories about Russia's alleged use of energy as a political weapon. A newspaper that attempts to report about any positive aspects of Russian energy geopolitics is unlikely to attract readers, no matter how carefully the story is researched. In Russia, conversely, the master narrative is that of the West seeking to destroy or, at best, misunderstanding Russia; pro-Western stories are subject to various forms of censorship, but above all they are simply not popular with Russian readers. We find the same phenomenon in the case of the recent surge of Chinese energy investments in Africa. In Western media, it has become almost impossible to report about this trend as something positive, while in Chinese media the reports cannot be negative. Any significant deviations from the mainstream narrative challenge the pre-established world views in the respective countries and regions, and hence they are repelled.[41]

So who actually controls world energy?

So who actually controls world energy? Do profit-hungry multinationals rule the world, or do governments and state-controlled companies have the ultimate say? To what extent do foreign-policy makers trump economic and environmental state actors? And are the key decisions taken in boardrooms and government offices in the world's capital cities, in the laboratories of corporate R&D departments in Beijing and Silicon Valley or in the mountain hideaways of terrorist leaders?

The only answer one can possibly provide to such impossible questions is that the geopolitics of energy is very much the outcome of complex interactions among the actors and actor categories discussed in this chapter. While the actors themselves play important roles, we need to analyse the intricate patterns through which they compete and cooperate with each other. Governments build alliances with each other to outmanoeuvre other governments in the quest for scarce global energy resources. Extractive companies forge international cartels to control international energy exports. The governments of the rich, import-dependent nations form united fronts, imposing import embargoes and preventing technology exports. Exporters, importers and transiteers compete for control over global flows of fuel and electricity. Military actors and defence ministers intervene. Fossil fuel companies struggle against nuclear power plant operators and against renewable energy entrepreneurs for market shares and profits. Companies lobby governments and international organizations. Inventors launch new radical technologies and try to attract investors. Courts decide to put a ban on one or the other energy source or project. Foreign offices quarrel with economy ministries over governmental decisions to be made. Environmental organizations forge ties with indigenous peoples against profit-hungry businesses. And the general public follows the exciting drama of energy and geopolitics through the media, while taking for granted that their own energy needs – in the form of electric lights, hot water, petrol – will continue to be met.

Exercises

- Who controls *your* energy supply? Who controls the energy system in your town/city/region/country? Have the patterns of control changed over time? List the actors that currently appear to be the most powerful.
- Pick a country and collect information about where, in the state administration, international energy issues are dealt with. Can you identify any particularly powerful individuals?
- Trace the history of a state-owned energy company other than the ones mentioned in this chapter. Take into account, in particular, its international activities. Has the company's ownership changed over time, and if so why? Who appears to be in control of the company at the present time?

Notes

1 Cf. Bridge, "Energy (In)security", 6; Dodds, *Geopolitics*.
2 See the websites of these companies.
3 Stent, *From Embargo to Ostpolitik*; Högselius, *Red Gas*.
4 Högselius, *Red Gas*.
5 See here, in particular, Millward, *Public and Private Enterprise*.
6 Shell, *Annual Report 2016*, 36, 46, 47.
7 Statoil, *Annual Report 2016*, 26.
8 See, for example, Stent, *From Embargo to Ostpolitik*, 17.
9 Högselius, *Red Gas*.
10 Yergin, *The Prize*, 592–632.
11 There are many alternative terms.
12 Stent, *From Embargo to Ostpolitik*.
13 For example, Högselius and Kaijser, *När folkhemselen blev internationell*.
14 Originally the PRC created a Ministry of Fuel Industry, which was split into a Ministry of Coal Industry and a Ministry of Petroleum Industry in 1955. See Peng, "The Evolution of China's Coal Institutions", and Högselius, "The Saudi Arabia of the Far East?"
15 Högselius, *Red Gas*.
16 Yergin, *The Prize*, Chapters 7 and 23; Millward, *Public and Private Enterprise*, 218–228; BP, "Our History".
17 Stern, *Future of Russian Gas and Gazprom*; Högselius, *Red Gas*.
18 Högselius, "The Internationalization of the European Electricity Industry"; Högselius and Kaijser, *När folkhemselen blev internationell*.
19 For a discussion see Noreng, "State-owned Oil Companies".
20 BP, *BP Statistical Review of World Energy 2017*, 38.
21 See the websites of these companies.
22 Högselius, *Red Gas*, Chapters 5–7.
23 See, for example, "Regeringen vill att Gotland säger nej till gasledningen", *Hela Gotland*, 13 December 2016; "Efter UD-mötet: så ska Gotland få ersattning för Slite hamn", *Hela Gotland*, 30 January 2017.
24 "Sinai Explodes into Violence after Years of Chronic Poverty and Alienation," *The Guardian*, 14 February 2012.
25 "Rosebud Sioux Tribe Calls House Keystone XL Passage an 'Act of War', Vows Legal Action", *Indian Country Today*, 17 November 2014.

26 For example, "Will Keystone XL Ever See the Light of Day?" *OilPrice.com*, 24 January 2018.
27 Dawson, *Eco-nationalism*; Högselius, "Connecting East and West?"
28 Alexander Nikitin, author interview, Oslo, 8 May 2007.
29 Mitchell, *Carbon Democracy.*
30 Yergin, *The Prize*, 629–630.
31 Molin, *Stenkol.*
32 Högselius, *Red Gas*, 192.
33 For example, Cantoni, *Oil Exploration, Diplomacy and Security in the Early Cold War.*
34 Ibid; Robert and Paglia, "Science as National Belonging".
35 Yergin, *The Prize.*
36 For example, Alter and Steinberg, "The Theory and Reality of the European Coal and Steel Community".
37 Fischer, *History of the International Atomic Energy Agency.*
38 Lagendijk, *Electrifying Europe*, 65.
39 Högselius et al., *Europe's Infrastructure Transition*, Chapter 2.
40 For example, Klare, *Rising Powers, Shrinking Planet*, 133.
41 Ciuta and Klinke, in the context of Western-Russian relations and German energy security, point to "an apparent dissociation between the media and governmental stance to the new Cold War and its embedded geopolitical logic". Ciuta and Klinke, "Lost in Conceptualization", 323.

Further reading

Dawson, Jane. *Eco-nationalism: Anti-nuclear Activism and National Identity in Russia, Lithuania, and Ukraine.* Durham, NC: Duke University Press, 1996.
Mitchell, Timothy. *Carbon Democracy: Political Power in the Age of Oil.* London: Verso, 2011.
Yergin, Daniel. *The Prize: The Epic Quest for Oil, Money and Power.* London: Simon and Schuster, 1991.

Bibliography

Alter, Karen J., and David Steinberg. "The Theory and Reality of the European Coal and Steel Community". Buffett Center for International and Comparative Studies, Working Paper No. 07-001, January 2007.
BP. *BP Statistical Review of World Energy 2017.* BP, 2017.
BP. "Our History". www.bp.com (accessed 18 May 2018).
Bridge, Gavin. "Energy (In)security: World-Making in an Age of Scarcity". *The Geographical Journal* 181, 4 (2015): 328–339.
Cantoni, Roberto. *Oil Exploration, Diplomacy and Security in the Early Cold War: The Enemy Underground.* Abingdon: Routledge, 2017.
Ciuta, Felix, and Ian Klinke. "Lost in Conceptualization: Reading the 'New Cold War' with Critical Geopolitics". *Political Geography* 29, 6 (2010): 323–332.
Dawson, Jane. *Eco-nationalism: Anti-nuclear Activism and National Identity in Russia, Lithuania, and Ukraine.* Durham, NC: Duke University Press, 1996.
Dodds, Klaus. *Geopolitics: A Very Short Introduction.* Oxford: Oxford University Press, 2007.
Fischer, David. *History of the International Atomic Energy Agency: The First Forty Years.* Vienna: IAEA, 1997.

Högselius, Per. "Connecting East and West? Electricity Systems in the Baltic Region". In *Networking Europe: Transnational Infrastructures and the Shaping of Europe, 1850–2000*, edited by Erik van der Vleuten and Arne Kaijser, 245–277. Sagamore Beach, MA: Science History Publications, 2006.

Högselius, Per. "The Internationalization of the European Electricity Industry: The Case of Vattenfall". *Utilities Policy* 17, 3 (2009): 258–266.

Högselius, Per. *Red Gas: Russia and the Origins of European Energy Dependence.* Basingstoke and New York: Palgrave Macmillan, 2013.

Högselius, Per. "The Saudi Arabia of the Far East? China's Rise and Fall as an Oil Exporter". *The Extractive Industries and Society* 2 (2015): 411–418.

Högselius, Per, and Arne Kaijser. *När folkhemselen blev internationell: elavregleringen i historiskt perspektiv.* Stockholm: SNS Förlag, 2007.

Högselius, Per, Arne Kaijser and Erik van der Vleuten. *Europe's Infrastructure Transition: Economy, War, Nature.* Basingstoke and New York: Palgrave Macmillan, 2016.

Klare, Michael T. *Rising Powers, Shrinking Planet: The New Geopolitics of Energy.* New York: Metropolitan Books, 2008.

Lagendijk, Vincent. *Electrifying Europe: The Power of Europe in the Construction of Electricity Networks.* Amsterdam: Aksant, 2008.

Millward, Robert. *Private and Public Enterprise in Europe: Energy, Telecommunications and Transport, 1830–1990.* Cambridge: Cambridge University Press, 2005.

Mitchell, Timothy. *Carbon Democracy: Political Power in the Age of Oil.* London: Verso books, 2011.

Molin, Harry. *Stenkol: En skrift utgiven med anledning av Svenska Stenkolimportörers Förenings verksamhet 1902–1952.* Stockholm, 1952.

Noreng, Øystein. "State-owned Oil Companies: Western Europe". In *State-owned Enterprise in the Western Economies,* edited by Raymond Vernon and Yair Aharoni, 133–144. London: Routledge, 1981.

Peng, Wuyuan. "The Evolution of China's Coal Institutions". Working Paper No. 86, Freeman Spogli Institute of International Studies, Stanford University, 2009.

Roberts, Peder, and Eric Paglia. "Science as National Belonging: The Construction of Svalbard as a Norwegian Space". *Social Studies of Science* 46, 6 (2016): 894–911.

Shell. *Annual Report 2016.* The Hague: Shell, 2017.

Statoil. *Annual Report 2016.* Oslo: Statoil, 2017.

Stent, Angela. *From Embargo to Ostpolitik: The Political Economy of West German-Soviet Relations, 1955–1980.* Cambridge: Cambridge University Press, 1981.

Stern, Jonathan. *The Future of Russian Gas and Gazprom.* Oxford: Oxford University Press, 2005.

Yergin, Daniel. *The Prize: The Epic Quest for Oil, Money and Power.* London: Simon and Schuster, 1991.

4   Understanding energy dependence

There was a time when virtually all energy in the world was consumed in the immediate vicinity of where it was produced. Not so anymore: energy supply has become global, or nearly so. How can we understand the mechanisms behind this pervasive transformation? Has energy's internationalization been an inevitable, unstoppable process, or is it rather the result of actors' deliberate strategies and initiatives? Why do actors engage in building international energy relations, and why do they sometimes refrain from doing so? And how has this changed over time? These are some of the basic questions to be explored in this chapter.

The central concept in the chapter will be that of energy dependence. We will explore the dependence theme from multiple angles. First, we will try to get an idea of "how international" the world's energy supply actually is, and look at the long-term patterns of change in international energy dependencies at an aggregate statistical level. We will then move beyond statistics and explore the underlying meanings of the dependence concept. We do this by scrutinizing a number of seemingly paradoxical cases of energy dependence, which can be unravelled by mobilizing our systems perspective. Third, we turn to a discussion about the underlying *motivations* that encourage actors to build – or not to build – relations with foreign actors. This translates into studying the opportunities and risks linked to energy dependence. Most of the chapter takes the perspective of import-dependent nations, but towards the end we will also turn to the phenomenon of energy *export* dependence. Finally, we discuss specifically how the rise of renewable energy sources is changing the global patterns of energy dependence, and how dependence is perhaps better understood within the broader notion of *interdependence*.

How international is the world's energy supply?

Few countries in the world can boast of being independent of energy supplies from abroad. Even by the crudest measure of energy dependence – net energy imports as a share of total domestic energy consumption – all 28 of the European Union's member states depend greatly on other countries for their energy supply, none of them having a positive energy trade balance. The EU as a whole imports 54% of its energy needs (with intra-EU energy flows being treated as "domestic").[1] The most advanced East and Southeast Asian nations – Japan, South Korea, Taiwan and Singapore – face an even more precarious situation, being almost totally dependent on imports to meet their energy needs.[2] Other countries are richly endowed in terms of domestic fuel resources. Even so, many of them – including the world's two most populous nations, China and India – import far more fuel than they export.

Finer measures reveal a more complex – and, from the perspective of importers, more troublesome – overall picture. Most obviously, countries that are net energy exporters at the aggregate statistical level are usually net importers of one or the other of the individual fuels they need. By the 1970s, for example, the Netherlands, thanks to vast discoveries of domestic natural gas, had become a net energy exporter, but the country remained critically dependent on imports for a range of other fuels, including crude oil – in a way that made it highly vulnerable to the turmoil of 1973/1974 and 1979.[3]

Moreover, if we move away from regarding energy sources as distinct commodities and, instead, adopt a systems perspective, it becomes clear that many countries are dependent on foreign nations not only for the supply of various fuels but also for a variety of technologies, processes and services relating to different parts of the energy system. In nuclear energy, for instance, energy dependence cannot be properly grasped by looking solely at uranium imports; dependence management here has much more to do with securing agreements with foreign countries for conversion, enrichment and spent fuel services. Similarly, in oil and gas, access to technology and equipment is key to system-building, a fact that has made even powerful fuel exporters – like Russia and Saudi Arabia – critically dependent on other countries in seeking to meet domestic demand and enable exports.[4]

Another way to quantify energy's internationalization is to look at the share of primary and secondary energy sources that, in the course of their journey through their respective systems, crosses one or more national borders. By this measure, it can be seen that *uranium* is by far the most internationalized fuel. Around 90% of global uranium production is exported from the extractive country to the country where the nuclear power plant is located. The latter countries comprise mainly EU member states (most of which import 100% of their uranium needs) and the United States (89% in 2016).[5] Uranium's internationalization is very high because several of the leading uranium-mining countries – notably Kazakhstan (which produced nearly 40% of global uranium in 2016), Australia and Niger – do not operate any nuclear power plants themselves.

From a wider systems perspective, uranium's internationalization becomes even more striking, since most nuclear energy countries do not possess mid-stream facilities such as conversion and enrichment plants. Especially in the smaller nuclearized countries – like Sweden, Finland and

Belgium – nuclear power plant owners usually find that they have to con-clude agreements with at least two foreign countries to make the fuel move from the mine to the reactor.

Oil, of course, is also highly internationalized. Approximately 48% of the world's oil is exported from the country of extraction.[6] The figure is lower than for uranium because some of the world's largest *oil-producing* nations – notably the United States, China and Russia – are also massive *oil consum-ers*. Thanks to the shale – the "tight oil" – revolution, as of 2018 the United States is the world's largest oil producer, but basically all of this oil is needed domestically so that the country remains a net importer. China, for its part, is the world's eighth largest oil producer, but nowadays virtually none of that oil is exported; the entire production is needed to meet domestic demand – and huge volumes of imported oil are needed in addition. Up to 1993, China was able to cover internal needs through domestic production, relying on "elephants" such as the famous Daqing field in Manchuria and the Shengli field in Shandong. Nowadays, the country imports around 70% of its oil.

Only 31% of global *natural gas* production was exported in 2016.[7] This is perhaps surprising in view of the heated debate about natural gas import dependencies, especially in Europe. Again, the aggregate world figure is distorted by the fact that several of the world's leading gas-producing na-tions use the lion's share of their production for domestic needs. Hence, Russia may be known as the world's largest gas exporter, but total Russian gas exports – 205 billion cubic metres (bcm) as of 2016 – are dwarfed by its own gas consumption, which measured 391 bcm.[8] Likewise, while American shale gas has been much discussed in terms of its impact on global energy markets, only 65 bcm, or 9% of US gas production, was ex-ported as of 2016.[9] Britain and the Netherlands have traditionally been two other large-scale gas producer-consumers, although their domestic reserves are now rapidly being depleted.

The dominance of large producers that are also large consumers is even more striking in the case of *coal*. When all coal types are considered, 18% of global production was exported in 2016. At a more detailed level, the corresponding figure for steam coal (for use in electric power plants) was 19%, for coking coal (for the steel industry) 29% and only 1% for lignite.[10] China, India, the United States and Russia dominate both global produc-tion and global consumption. As in the case of oil and gas, the far-reaching dependencies of the smaller nations on coal imports are thus diluted at the aggregate statistical level.

Biofuels and *electricity* are the least internationalized of the world's ma-jor energy sources, with only around 5–10% of global production being exported. As we will see later on, however, this may change in the future.

Finally, it is worth mentioning that, even in the twenty-first century, some energy sources are hardly internationalized at all. Examples include *firewood* and *charcoal*, which are of immense importance in many poorer parts of the world. In large parts of Asia and Africa, hundreds of millions of women spend much of their days collecting firewood, used especially for cooking. It is a tedious activity and there is no lack of private and public actors who want to liberate these families from it – and insert them into modern, globalized energy systems. The transformation is already happening – and it is being accompanied by intense debates and struggles for control.

Long-term patterns of change

Energy dependence is prone to change over time. To take the earliest example of international energy dependencies, starting in the nineteenth century a number of European nations became dependent on British – and subsequently on German – coal supplies. However, British coal production peaked in 1913 and German production in 1956. Since then, both nations have seen their net coal exports dwindle and turn into net coal imports. By the 1990s, what had once been the world's two most powerful coal exporters had become Europe's largest coal importers! By 2017, however, Britain had managed to virtually eradicate its coal import dependence through a shift away from coal to other fuels in electricity production, and the downsizing of its steel industry. Germany, by contrast, still remains heavily dependent on coal imports.

The United States, for its part, totally dominated the world oil market up to the interwar era, and numerous countries worldwide became critically dependent on the US-based oil majors for their supplies. By 1949, however, the United States had become a net oil importer, and when the first oil crisis erupted in 1973, Americans were shocked to find their consumption strongly impacted by the political machinations in the Middle East. But several technological revolutions in the oil industry subsequently changed the pattern once again. New advances in geophysics, deep-sea exploration, and above all, hydraulic fracturing and horizontal drilling, enabled American oil producers to mobilize new domestic petroleum resources – paving the way for a new era of American near-independence in oil.

Denmark, a much smaller country, has gone through a different process. In the course of the post-war decades, it became almost totally dependent on oil from the Middle East. By the late 1970s, however, it had managed to bring into production its own offshore oil field in the North Sea; as a consequence, its oil import dependence was radically reduced as ever-larger shares of its oil needs were covered by this domestic production. The process was helped by the fact that absolute oil consumption was reduced. By 1997, Denmark had become a net oil exporter, being totally independent, in statistical terms, of foreign supplies. Subsequently, however, Danish oil production peaked in 2005. A decade later, production was only a third of what it had been in the early 2000s. As a result, the country once again became a net oil importer.[11]

The situation is more extreme in a range of developing countries. Egypt, for example, has historically played an important role in the international oil system through its control of the Suez Canal and the Egyptian government's agency in the 1956 Suez Crisis, the crisis that accompanied the 1967 Six-Day War and, in particular, the 1973 Yom Kippur War. Yet the country was not a major oil producer, and up to 1967 it was a net oil importer. Starting in the mid-1970s, however, domestic Egyptian oil production experienced a breakthrough and grew rapidly before peaking in 1993. During this period, the country was able to export over half of its production and did not need to worry about energy oil import dependence. Its role as an exporter has not lasted, however. In the course of the past quarter-century, domestic Egyptian oil consumption nearly doubled, while its oil industry was unable to further boost production. As a result, since 2010, Egypt, for the first time since 1967, has once again become dependent on oil imports.[12]

These examples illustrate how a range of factors interact to produce energy import dependencies at an aggregate statistical level. Apart from domestic fuel endowments, domestic consumption is clearly a key factor in shaping energy dependency relationships. In the world's richest countries, energy consumption is no longer growing; on the contrary, it is often shrinking. It is still climbing, however, in many developing countries, both small and large, in a way that strongly increases their energy dependence. In a few lucky countries, growth in domestic consumption has been matched by scaled-up domestic production – which has prevented import dependencies from materializing. Yet even in these cases, the very nature of fossil fuels as finite, non-renewable energy sources has made such "golden eras" of independence all too brief. In the case of Egypt it lasted four decades, in the case of Britain, a quarter-century, and in Denmark, less than two decades (as far as oil is concerned). The only thing that can put an end to energy import dependence once and for all is, clearly, totally phasing out the fuel in question, as in Italy's decommissioning of its nuclear energy plants after the 1986 Chernobyl disaster, or the ongoing phase-out of coal use in the British electricity system. Which country will be the first to phase out oil in the same manner? That remains to be seen.

Paradoxical dependency relations

Statistical indicators of cross-border flows of fuel tell us only part of the story of energy dependency relationships. This is because fuel importers are in reality primarily dependent not *on the fuel* as such, but rather *on the system* through which the fuel is supplied. This has interesting and far-reaching consequences for our analysis, some of which may appear paradoxical at first glance.

For example, while many EU countries are highly dependent on Russian natural gas, our systems perspective forces us to acknowledge that this is a much too simplified view when analysing EU-Russian natural gas relations. An interesting component of these relations has to do with the role of gas storage facilities in the interlinked EU-Russian gas supply system. One of the largest European storage facilities is located just outside Riga in Latvia.

It was constructed during the Cold War, at which time Latvia was part of the Soviet Union. The facility was built to counter seasonal supply problems, especially in winter, throughout the Baltic region and also in St. Petersburg (then Leningrad). Since the collapse of the Soviet Union, it has continued to play an important role far beyond Latvia itself. Estonia and Lithuania draw heavily on the facility during the winter, and so does Russia, which thus, paradoxically, at times depends on Latvia for its gas supply security.[13] Finland has recently sought access to the Latvian storage facility as well, hoping to link up with the Baltic transmission system through a submarine pipeline under the Gulf of Finland. Linking up with Latvia is considered a useful way for Finland to strengthen its supply security – in spite of the fact that Latvia does not possess any domestic gas deposits at all![14]

Another dependency that comes to the fore here has to do with the critical "choke points" in the global oil supply system. For example, around 20% of the world's oil production has to pass through the Strait of Hormuz in order to reach markets, and as a result a closure of this strait would have disastrous consequences for the world's energy supply. The same goes for the Strait of Malacca and, to a lesser but still significant extent, the Suez Canal, the Turkish Straits, the Danish Straits and the Panama Canal. The US Navy plays an active role in keeping several of these critical choke points and waterways open. As a result, many of the world's oil-importing nations have become dependent on the United States for their oil supply security – even though they do not import a drop of oil from America. The oil-exporting nations also depend on the US Navy's protection – and, in the case of Russia, on the Turkish and Danish state agencies that watch over

Figure 4.1 Inside the gas storage facility at Incukalns, Latvia. A specialist works in the main control room of Latvijas Gaze's underground storage facility, August 2014. Photo: Ints Kalnins/Reuters/TT.

Russian oil tankers on their way from Baltic and Black Sea ports to foreign markets. (The official status of the Bosporus as an international waterway does not prevent the Russians from worrying about the possible consequences, for Russian oil exports, of an international conflict in the area.)

Another revealing case is Iran's dependence on petrol imports. Iran has long been one of the world's largest crude oil producers, and it might thus seem paradoxical that the country is one of the world's most import-dependent nations when it comes to petrol. The mismatch is, unsurprisingly, explained by the lack of domestic refinery capacity in Iran. For decades, the Iranian government subsidized domestic petrol sales very heavily in order to boost modern automotive society and keep car owners and other petrol users happy. Combined with economic growth and a rapid population increase, this soon generated domestic petrol shortages. Refiners were not able to keep pace with the skyrocketing demand. The situation was worsened by the emergence of a lucrative smuggling business, with smugglers selling subsidized petrol at much higher prices abroad, thus further worsening domestic shortages. Iran turned to imports to fill the gap between domestic supply and demand, but the dependencies that this generated were soon identified by state actors as a political risk. In response, President Ahmadinejad's cabinet launched a radical "Gasoline Rationing Plan" in 2007, seeking to curb domestic petrol (gasoline) consumption. The plan was based on the fear that the country's petrol import dependence might be used by Western powers in putting pressure on Iran to abandon its controversial nuclear technology programme.[15]

A further paradoxical example is Morocco's dependence on German electricity. Morocco does not import any electricity from Germany, yet from a systems perspective, it depends on that country, because they are part of the same synchronized electricity grid. As a matter of fact, in this case, 24 (mainly European) nations depend on each other in highly critical ways, even though electricity is traded between them only on the margin. The synchronized connections have been established mainly to strengthen overall grid stability, in such a way that the reliance on expensive domestic back-up power plants can be minimized and largely replaced by deliveries from other grid members in case of crisis. Yet the interconnections also come with the risk that a major failure in one national grid may spill over to neighbouring countries. On 4 November 2006 precisely this happened – on a grand scale – as a power failure in northern Germany cascaded all the way down to the Iberian Peninsula and onwards across the Strait of Gibraltar to Morocco. There, schoolchildren doing their homework under the evening lamp found themselves suddenly in the dark because of the German failure thousands of kilometres away in northern Europe.[16]

The most telling case of energy dependence from a systems perspective, however, is found in the field of nuclear energy. Take Finland: it operates four nuclear reactors (a fifth is being built), two of which have traditionally relied on Russian uranium and the other two on Western supplies. But the statistical figure that specifies the volume of uranium imports and the geographical sources of that uranium do not tell us much about Finland's import dependence on nuclear fuel. As of the late 1970s, the "Western" reactors relied on an exceedingly complex geographical route for their fuel supplies. TVO, the operator, purchased uranium extracted from Canadian mines. From there, the ore was shipped across the Atlantic to La Hague on the French coast for conversion into uranium hexafluoride. After that, the fuel embarked on another lengthy voyage, heading for the port of Riga in Soviet Latvia, where it was reloaded onto railway cars for transport to the Mayak nuclear complex in the eastern Urals. Enriched in the Soviet Union, the fuel reappeared a few weeks later in Leningrad. Next, it floated across the Baltic to the fuel element factory at Västerås, Sweden, after which it could eventually be transported, in its final form, to TVO's nuclear power plant at Olkiluoto on the Finnish coast. All in all, TVO thus depended on no fewer than four countries – Canada, France, the Soviet Union and Sweden – for the reactor's fuel supply. For each step in the supply chain, TVO concluded a detailed contract with the respective supplier.[17]

BOX 4.1 IS NUCLEAR ENERGY A DOMESTIC ENERGY SOURCE?

According to official EU statistics, the EU area as a whole covers 46% of its energy needs from internally available energy sources and 54% from external (non-EU) sources. At the level of individual EU member states, aggregate energy import dependencies vary widely, from over 70% for Cyprus, Malta, Luxembourg, Ireland, Belgium, Lithuania, Portugal, Italy, Spain and Greece, to less than 40% for Estonia, Denmark, Romania, Poland, Sweden, the Czech Republic, Bulgaria and the United Kingdom. Germany, Austria, Slovakia, Hungary, the Netherlands, Latvia, Slovenia, Croatia, Finland and France are somewhere in between.

However, there is – arguably – at least one fundamental problem with these figures: they count nuclear energy as a domestic energy source. This is a bold and intriguing claim. After all, the Czech Republic is the only EU member state that actually extracts uranium ore, and deliveries from that source cover only 1.5% of the EU's total uranium needs. The rest, 98.5%, is imported, mainly from Niger (22% of total needs – this goes mainly into French reactors), Canada (20.6%), Russia (19.3%), Kazakhstan (15.8%), Australia (13.2%) and Namibia (3.5%). A few years ago, Romania and Bulgaria also produced some uranium, but those mines have now been shut down. So how do these uranium supply patterns make nuclear energy an intra-European energy source?

Figure 4.2 The Tamgak open air uranium mine, seen at Areva's Somair uranium mining facility in Arlit, Niger, September 2013. The Sahara Desert is the most important source of uranium for European nuclear energy production. Photo: Joe Penney/Reuters/TT.

One possible, but not very convincing, answer is that while the EU imports virtually all of its uranium ore, it relies to a much greater extent on internal supply when it comes to *enrichment services*. As of 2016, 70% of the uranium needed in European nuclear power plants was enriched in the EU, either by Areva in France or by Urenco in Germany, the Netherlands and Britain (which at the time was still an EU member). Most of the remainder, 28%, was enriched in Siberia by the large Russian nuclear fuel company TVEL. However, if we let energy dependence be defined by the level of internal enrichment rather than by the share of internal uranium ore mining, we would need to do the same in the case of, say, oil. As a result, we would have to state that the EU is self-sufficient in oil, because basically all crude oil consumed in the EU is processed by EU-internal oil refineries. Yet there is a strong consensus among oil analysts that EU member states are import-dependent in oil as long as the crude oil comes from non-EU sources. So this answer is inconsistent at best.

Another possible, and somewhat more convincing, answer relates to the structure of *energy storage* in nuclear power. The nuclear industry stores nuclear fuel in much larger volumes than the coal, oil or gas industries. As of 31 December 2016, there were around 51,000 tonnes of natural uranium or the equivalent in the inventories of EU nuclear operators. Some of it was in the form of natural uranium, while some had already undergone refining (conversion and/or en-richment) or been turned into ready-to-use fuel elements. Total EU uranium reactor requirements amount to about 16,000 tonnes per year. Hence, the uranium kept in inventories covered the total needs of the EU's 128 nuclear reactors for the next three years! The implica-tion, from an energy security point of view, was that even in the case of a total supply disruption of uranium from EU-external sources, Europe would be able to continue producing copious amounts of nu-clear electricity – for longer than anybody could imagine that such a supply crisis would ever last.

The extremely large inventories of nuclear fuel are probably the main reason why EU statisticians insist on labelling nuclear energy a "domestic" energy source. But clearly, this claim is subject to a great deal of interpretative flexibility.[18]

Technical and chemical aspects

Another way to assess energy dependence relationships is to study the ex-tent to which it is possible for an importer to switch to an alternative sup-plier. A common argument in the debate about oil import vulnerability is that a potential disruption in supply from one source can easily be compen-sated for by bringing in supplies from another source. In fact, refineries are often locked into one particular type of crude oil, so that it becomes much more difficult – and costly – than envisaged to actually switch to a totally new source of supply. For example, over the course of the 1970s, several American refineries had adapted their processes to the specific characteris-tics of Iranian light crude. Following the Iranian revolution of 1978/1979, these supplies were disrupted. The refiners looked for alternative supplies but did not find sufficient volumes of crude oil of the same chemical qual-ity. Only heavier crudes were available, which the refineries were unable to process. As a result, a number of them came to a standstill. The upshot was a gasoline shortage in the United States.[19]

More recently, the US oil industry has faced the opposite problem. Most American refineries are nowadays geared to a blend of light and heavy crudes. They cannot operate on light crudes only. This became a problem in the 2010s as Washington eyed domestic shale oils as a pathway to US oil independence. Most US shale oils are very light, so as shale production increased, American refiners were able to gradually phase out imports of foreign light crudes. By 2014, imports of Nigerian and some other light foreign oils had been slashed, while a domestic surplus of light shale oils emerged. But many refiners remained dependent on imports of the heavier crudes that were also direly needed. Domestic shale oil producers were not able to supply these. Most refiners did not consider this to be much of a problem, since it was easy for them to source heavier crudes on the international oil market. The shale oil producers were also quite happy, eyeing profitable exports of the surplus that US refiners were not able to absorb. Washington, however, did not agree, pointing to the ban on US oil exports that had been in place since 1975. The ban was to remain effective as long as the country was not self-sufficient in oil, and this was not yet the case. As late as 2014 President Obama's Secretary of Energy Ernest Moniz insisted that the export ban remain in place. The implication was that numerous refiners would have to rebuild their refineries – at great cost – in such a way as to adjust to the lighter shale oils. This was hardly economically rational, although it seemed to make sense from an "energy security" point of view. In December 2015, however, Washington changed its mind and lifted the oil export ban. As a result, the United States is now both a major exporter and a major importer of crude oil; as of spring 2018 the country imported 7.6 million barrels of crude oils per day while exporting between 1.5 and 2 million barrels.[20]

Brazil's experience offers a variation on the same theme. In the post-war era, Brazil's known oil deposits were very limited. Like many other countries, Brazil turned to supplies from the Middle East to cover its needs, becoming heavily dependent on light crudes from that region. In the 1960s and 1970s, several Brazilian refineries were built to process these imports. Later on, vast domestic oil deposits were discovered in Brazil itself, allowing the country to emerge as a net oil exporter from 2011. However, most of the domestic crudes were heavy ones. For this reason, the domestic oil industry opted to maintain a large part of its oil imports. It would be extremely expensive to reconstruct refineries in such a way as to adapt them for the heavier domestic oil.

BOX 4.2 JAPAN'S DEPENDENCE ON CHINESE CRUDE OIL

There was a time when Japan depended on China for parts of its crude oil supply. In the 1950s, Chinese oil explorers, in cooperation with Soviet experts, discovered vast oil deposits on Chinese territory. By the 1970s, the available resources had grown so large that an exportable surplus was emerging, and Western industry journals were even speculating that China was on its way to become a "Saudi Arabia of the Far East". At the same time, Japan was becoming painfully aware of its over-dependence on Middle Eastern oil and was looking for opportunities to diversify its imports in geographical terms. From a technical-economic point of view, there was thus good reason for both China and Japan to contemplate a cross-border oil-trade arrangement. A problem was that the two nations were still officially at war with each other (no peace treaty had been concluded after the end of the Sino-Japanese War in 1945), so the situation was somewhat murky from a political point of view. A number of foreign policy actors opposed the idea of Chinese exports to Japan. But in 1978 the Chinese and Japanese governments concluded a "peace and friendship agreement" that formally put an end to the war. Intimately linked to that agreement, a five-year oil deal was signed, foreseeing exports of no less than 348 million barrels of oil from 1978 to 1982. The Japanese oil industry welcomed this new source of supply. The only problem was that the crude came from China's famous Daqing oil field, located in the country's northeast. Daqing oil was widely known to be extraordinarily heavy. This forced the Japanese to re-design their refineries, adjusting the plants to the special character-istics of Daqing crude. As a result, a significant part of the Japanese oil industry became locked into supplies from China. This became a problem for Japan when the Chinese, in 1980, told their Japanese partners that they faced problems in scaling up oil production and that they, as a result, would not be able to deliver more than roughly half of the promised supplies originally foreseen for the year 1982. Having adjusted to Daqing's heavy crude, it became an extremely expensive operation for the Japanese refineries to refit in order to be able to process supplies from elsewhere.[21]

There are many similar cases that make clear how energy dependence, seen from this chemical-technical systems perspective, goes far beyond what can be conveyed through simple statistical figures. Steel industrialists, for example, cannot make use of just any type of coal, but are critically dependent on high-quality coking coals. Only such high-quality coals can ensure that the steel becomes strong enough – and the more advanced the steel, the more dependent the mill is on suitable fuel.

Another notable case is Europe's natural gas system. This system is highly integrated, and gas is blended from several different sources. From a secu-rity of supply point of view, this is clearly advantageous, since in the case of supply problems from one direction, it is possible for gas from elsewhere to come to the rescue. A closer look, however, reveals that the European gas system was built not as one, but as two systems: an "H-gas" and an "L-gas" system, each one featuring a different standard in terms of the calorific value of the gas. The vast Groningen gas field in the Netherlands is the main source of L-gas, whereas the North Sea, Algeria, Russia and a few other sources supply H-gas. The calorific difference means that Dutch gas is not interchangeable with Russian or Algerian gas, and hence cannot compensate for any supply disruptions from these sources. Only techni-cally complex and economically expensive conversion facilities can enable the two systems to interact in physical terms. Indeed, such facilities have been built in a few places, but critics complain that they make no economic sense.[22]

Nuclear fuel dependencies are also more complex than they seem at first glance. We have already seen that while almost all nuclear power plant operators in the world draw on foreign supplies of uranium ore, the key dependencies are linked to refined rather than crude forms of uranium. Conversion and enrichment services, which are controlled by only six or seven companies worldwide, were mentioned in this context. Further de-pendencies arise from fuel fabrication. TVEL, the Russian nuclear fuel agency, traditionally holds a *de facto* monopoly on fuel fabrication for all water-water energetic reactors (VVER), in both Russia and abroad. Such reactors, designed and built during the Soviet era, are currently operat-ing in Ukraine as well as in five EU member states (Finland, the Czech Republic, Slovakia, Hungary and Bulgaria). Following the Cold War's end, the EU (through its Euratom Supply Agency) and US state agencies, in close cooperation with local governments and authorities, initiated attempts to diversify nuclear fuel supplies for the Soviet-built reactors in order to reduce Russian control over the fuel cycle. Westinghouse, in particular, became engaged in developing VVER-compatible fuels. The company entered the market by offering attractive prices (partly with the help of EU and US government subsidies). However, most of the nu-clear power plant operators did not show much enthusiasm, preferring to continue their long-standing cooperation with the Russians, whose VVER-related expertise was perceived as superior to Westinghouse's. The preference for the Russians seems to have been rooted in an ap-preciation of the fuel's technical quality as totally decisive for the safe and economic operation of the reactors. Seemingly minor deviations from top quality could potentially force unplanned shutdowns of en-tire reactors, with substantial negative effects on electricity sales and grid stability. The Czech Republic's Temelin NPP, for example, which gave the Westinghouse option a chance in the early 2000s, experienced "massive malfunctions related to the geometric stability of the fuel that eventually led to premature unloading of all of Westinghouse's fuel as-semblies". Against this background, VVER operators have generally not been willing to buy Westinghouse's fuel, even though it has sometimes been cheaper than TVEL's.[23] The main exception to this day is Ukraine's decision to switch from TVEL to Westinghouse as its main nuclear fuel supplier, motivated by Russia's 2014 annexation of the Crimea and the ensuing war in eastern Ukraine. Radical geopolitical developments thus changed the Ukrainian nuclear industry's preparedness to accept the technical and economic risks of using non-Russian fuel.[24]

Dependence as opportunity

Has energy's internationalization been an inevitable, unstoppable process, or is it rather the result of actors' deliberate strategies and initiatives?

From the mid-nineteenth century onwards, the rise of modern transport infrastructures – railways, steamships, mechanized ports, paved roads, trucks, etc. – made it possible to ship large volumes of fuel around the world at astonishingly low cost. In combination with the "divine laws of supply and demand" – as Saudi oil minister Yamani once put it[25] – this has sufficed to stimulate a voluminous international trade in everything from coal and oil to uranium and biofuels. Dedicated energy infrastructures – supertankers, oil and gas pipelines, high-voltage transmission lines and so on – have further increased the economic attractiveness of long-distance fuel movements. Political conflicts and trade barriers have worked against this logic at times, but in general there have been very strong structural pressures pointing to the globalization of energy as something very natural. From this point of view, energy's internationalization may actually appear as something nearly inevitable.

But at the same time, it all hinges on the actors. The individuals and organizations involved in making the world of energy go around are not machines. They are not pre-programmed. They do not necessarily follow any rational algorithms. They are humans: men and women with their own specific world views, agendas, visions, moods and desires. They engage in decision-making processes whose outcomes cannot be predicted, choosing among alternative avenues of action, between the options that they think are available. And they interact with each other, so that the end result, more often than not, is a compromise, a synthesis of multiple actors' ideas, wills, opinions and proposals. So from this point of view, energy's internationalization is more the outcome of actors' individual behaviour and the complex social processes in which they engage.

But *why*, then, do actors engage in building international energy relations? It is possible to discern at least four sets of "drivers", each of which is linked to a particular kind of *opportunity*. These opportunities are discussed in the following, whereby for the moment we delimit our analysis to the perspective of import-dependent nations.

First, actors often engage in or promote dependency relations because they spot *supply opportunities*. These imply that actors take an interest in foreign supplies of energy for the simple reason that domestic supplies are non-existent, or not available in desired quantities. In pre-modern times, such lack of fuel was unusual, as locally sourced firewood constituted by far the most important energy source. Few human communities faced a total lack of nearby forests, and although some were better endowed than others, there was a certain degree of evenness in fuel availability. The rise of coal and oil as new pervasive sources changed the situation. Some countries entirely lacked local deposits of one or both of these basic fuels, and many others needed larger volumes than were available domestically. This had far-reaching consequences, especially in view of the fact that both coal and later oil became totally indispensable for countries with ambitions to industrialize. Industrial engines, locomotives, steamships and urban gasworks were difficult or even impossible to operate on the basis of firewood, and in the twentieth century several new means of transport and industrial production became heavily dependent on oil. As a result, for countries without domestic coal and oil deposits, the conclusion of import agreements with foreign suppliers of coal and oil became an existential component of their industrialization and modernization strategies. Not exploiting such supply opportunities would have left them far behind in terms of economic and social development.

Natural gas and uranium have not been considered indispensable to the same extent as coal and oil. Some of the world's most advanced economies, like Norway and (most of) Sweden, have managed to do well virtually without natural gas as a component of their domestic energy systems (although Norway, ironically, is one of the world's largest exporters of natural gas). Similarly, only a limited number of countries worldwide have chosen to exploit the supply opportunities linked to uranium as a fuel. The rest has done well without nuclear power plants. Yet most countries in the world that operate natural gas systems and nuclear power plants do so on the basis of imported fuels, and in most cases the gas and uranium needs are so large that it would be physically impossible to cover the domestic needs from local sources. Only in a few very large and well-endowed nations, notably Russia and Canada, may a supply strategy entirely excluding imports be realistic.

Actors have also taken great interest in the supply opportunities linked to imports of *refined* forms of energy. Many countries lack the technical competence, the financial strength and/or the facilities needed to convert and refine some fuels, but thanks to imports they have been able to compensate for this. Even countries that are richly endowed with crude forms of energy have often taken great interest in the supply opportunities linked to refined energy. As we have seen, Iran found it useful to import large volumes of petrol when domestic oil refineries proved unable to meet domestic needs. Luxembourg, for its part, is so small that it has found it unrealistic to construct its own domestic oil refineries; instead, it conveniently exploits the supply opportunities of imports from Belgium and, to a lesser extent, Germany, France and the Netherlands. (The refined products from Belgium are imported through a pipeline from Antwerp that passes through German territory.)[26] In the nuclear field, many countries initially struggled with the lack of domestic uranium enrichment capacities. Constructing enrichment facilities was found to be so technically demanding that a number of nations with nuclear ambitions set out to develop special types of nuclear reactors that would be able to run on natural, unenriched uranium – heavy-water reactors. Such strategies seemed reasonable not least in view of the lack of internationally available enrichment services. In the mid-1950s, however, political decisions in the United States and the Soviet Union allowed countries with nuclear ambitions to conclude contracts with US and Soviet suppliers of enrichment services, thus generating radically new supply opportunities. This became a key motivation for a range of smaller nations with nuclear-energy ambitions to abandon their heavy-water projects.[27]

Secondly, actors see *economic opportunities* in linking up with foreign suppliers. This implies that international linkages are pursued as a way to reduce costs. Unsurprisingly, actors often prefer imported fuels over domestic ones if the former are cheaper – though this might give rise to objections on both security and employment grounds. The history of US oil imports illustrates this point. In the interwar years, consumers, independent refiners and energy-intensive industries in the United States were happy to notice that imported oil was becoming available at prices lower than those offered by domestic oil producers. They spotted economic opportunities in shifting away from domestic supplies to imports. American oil producers with their production basis in the United States itself, understandably, did not appreciate this foreign competition. Neither did the nation's oil workers. Further supported by political actors in oil-producing regions such as Texas and Oklahoma, they lobbied Washington to impose tariffs, quotas and bans on oil imports. Their interests here overlapped with those of the Pentagon, which saw oil imports as a potential security risk.[28] The economic opportunities linked to Western European imports of Soviet oil during the Cold War were also considerable, but likewise gave rise to controversy.[29] A similar development characterizes the history of the European coal industry, where the economic opportunities presented by imports have been pervasive. Over the past 20 years, the production of hard coal has been phased out in all Western European countries – in spite of the fact that there is still a lot of coal in the ground. However, power plants and steelworks continue to burn copious volumes of coal. The difference is that the coal is now imported. The reason for the transition is simple: the imported coal is cheaper.

Historically, economic opportunities of this kind have often arisen due to the lack of suitable domestic transport infrastructures. In such cases, it is actually not so much that foreign supplies are cheap, but rather that domestic supplies are overly expensive for the end user. Pre-revolutionary Russia, for example, did not lack domestic coal reserves, but the lack of a suitable domestic railway infrastructure raised the costs of supplying a city such as St. Petersburg with domestic coal. As a result, coal users in the Imperial capital opted to buy their coal from British suppliers.[30] Many coastal regions in China and India are currently in a similar situation. Both countries are immense coal producers, but even if domestic supply and demand roughly match, coal users along the coasts often find it more advantageous to draw on imports, given bottlenecks and inefficiencies in domestic coal shipments. However, (domestic and international) railway and road-building projects, along with pipelines and electricity transmission lines, may radically alter the relative economic attractiveness of domestic versus imported energy. The recent widening of the Panama Canal, for example, has suddenly made it economically attractive for East Asian fuel importers to conclude import agreements with US shale gas suppliers.

Another type of economic opportunity that has motivated energy companies to create cross-border connections is linked to emergency supplies. This is particularly evident in electricity. Especially in Europe, electricity companies have invested heavily in constructing interconnections with neighbouring countries. The underlying motivation has little to do with any large-scale prospective electricity trade. Although such trade does occur – as in the case of French electricity exports to Italy – the principal logic behind electrical cross-border integration has to do with its potential to increase network stability and with the prospects for foreign production to come to the rescue in case of domestic problems. Grid stability may also be strengthened without international connections, notably through construction (and continuous maintenance!) of massive reserve power plants, but operators usually view international integration as a more favourable option – because it is less expensive. The same logic is evident in the oil and gas industries, where storage mechanisms are increasingly becoming internationalized. An interesting example is the Finnish attempts to link up with Latvia's huge gas storage facility mentioned above; here, access to foreign storage capacity is viewed as economically much more favourable by the Finns than the alternative of a purely domestic storage solution.

Third, actors might view access to foreign energy supplies as an *environmental opportunity*. This implies that international linkages are pursued as a way of solving environmental problems. Political representatives, environmental organizations and the general public have often considered domestic extraction of fossil fuels, uranium and biofuel feedstocks as environmentally hazardous. Imports of the same fuels from abroad have then commonly been viewed as a way to cope with this problem. Of course, the substitution of imported for domestic energy does not necessarily reduce environmental impact in any absolute sense, as the problems are merely transferred to other countries. Locally, however, the environmental benefits may be substantial. This appears to be a major motivation, for example, for several countries in Europe, notably Germany, to rely on imported rather than domestic coal, and for others to import uranium from abroad for their nuclear power plants in spite of large domestic uranium reserves. Bulgaria, for example, has recently closed down its sizeable uranium mines due to environmental concerns. Instead, the country's nuclear power plants now run on imported uranium. Sweden is another highly nuclearized country that, in spite of sizeable domestic uranium reserves, relies exclusively on imports for its nuclear fuel supplies, the motivation being a combination of economic and environmental opportunities. Sweden's right-wing parties have argued that Swedish nuclear fuel supply security would benefit from domestic uranium mining, and the mining industry has recently showed increasing interest. But in May 2018, the Swedish Parliament voted to institute a total ban on uranium prospecting and mining in the country – for environmental reasons.[31]

Fourth and finally, actors have identified *political opportunities* in energy's internationalization. Such opportunities may take different forms, and they will be discussed in further depth in Chapters 6 and 7. Suffice it here to note that state actors have often viewed energy as a useful instrument of foreign policy – in both negative and positive ways. The notion of energy as a foreign policy tool is most commonly associated with the Organization of the Petroleum Exporting Countries (OPEC)'s "oil weapon" or Russia's "gas weapon", an intriguing and controversial conceptualization that discursively places energy policymaking on a par with physical violence and military action. Yet, as we shall see in Chapter 7, energy has historically also inspired attempts to *solve* international conflicts, build world peace and stimulate regional integration. Both aspects play substantial roles in shaping the geopolitics of energy.

Dependence as risk

The opportunities discussed in the preceding section explain why actors think it attractive to link up with foreign energy sources and systems. But why do they sometimes *refrain* from building such international links – that is, in spite of the opportunities that such links entail? To understand this, we must try and discern the variety of *risks* involved in energy's internationalization.

The rise of the international energy trade has generated a diversity of risks. The most obvious ones are what we may label *supply risks*, which can be defined as a fear of being left without energy in a material sense. In the pre-modern era, the fear of getting physically cut off from fuel supplies was rarely an issue, as most of the fuel by far – primarily in the form of firewood – was supplied locally. The main exceptions were river-adjacent industrial sites that relied on deliveries of driftwood from upstream sources, or coastal localities that received firewood shipments from foreign shores. The subsequent transition to coal generated supply risks at a totally new level. Even back in the nineteenth century, fears of being cut off from foreign coal supplies were already omnipresent in many import-dependent nations. As of 1900, Sweden's coal-import dependence was subject to heated parliamentary debates, with some MPs pointing to the rapidly rising coal imports as a "Damocles sword" hanging over the country.[32] Labour strikes in Britain's coal-mining regions and supply disruptions due to bad weather usually defined these early coal fears. The outbreak of World War I, which led to a collapse in the international coal trade, served to further emphasize the supply risks. Several European cities and industrial regions were cut off from much of their supplies and forced to ration the rest. At stake was not only coal use as such but also the production of secondary energy sources on its basis, such as electricity and town gas. As a result, factories stopped working, and many cities went cold and dark, while old lighting fuels such as vegetable oil made a comeback.[33] This crisis was repeated during World War II, by which time oil had also become a fuel of major civil importance, particularly in the transport sector.

Military actors were also aware of what the transition to coal and later oil implied. In the nineteenth century, the navies of many smaller nations faced the choice between remaining in the age of sailing or investing in new warships powered by imported coal. The US Navy was also preoccupied with the issue of coal supply risks, especially when it came to operations in foreign waters (such as the Mediterranean) and in American colonies.[34] Britain enjoyed the advantage of controlling the world's best steamer coal, extracted from Welsh mines, and its supply risks were thus less pronounced, at least as far as the Royal Navy's European operations were concerned. This changed in the 1910s as the Royal Navy switched from coal- to oil-powered warships. As we have seen earlier, this transition led to British import dependence on Persian oil. Supplies were recognized as risky, in spite of a strong British military presence in the Persian Gulf and British control of the Suez Canal.

Supply risks relating to transport – or transit – later became very pronounced in the geopolitics of energy. This is especially evidenced by well-known features in global energy supply such as the critical "choke points" through which much of the world's oil must pass on its way from the oil field to end users (see discussion above), and the politically "problematic" transit countries – notably Ukraine – that Russian natural gas traditionally passed through en route to Western Europe. Overall, transit-related supply risks have increased over the years through the growing reliance on grid-based energy systems, such as transmission lines for electricity and pipelines for oil and natural gas. Whereas transit routes for coal and oil can often be changed flexibly and on short notice – that was how the oil-importing countries and the international oil companies dealt with the 1956/1957 and 1967 supply crises – this has rarely been possible in the case of the disruption of a major submarine electricity cable or an international natural gas pipeline.

Second, actors discern *economic risks* in international energy relations. These are linked to the common fear among actors that imported energy will become overly expensive, especially in the context of cartelization attempts in and among the exporting nations. Even back in the nineteenth century, coal importers feared that exporters might distort the international market by forging cartels and manipulating prices. The Rhenish-Westphalian Coal Syndicate in Germany is a case in point; it was formed by the Ruhr coal producers in 1893 in an attempt to keep prices high, and since a large portion of their total production was exported, this gave rise to fears not only in Germany itself but also elsewhere in Europe (notably in the Netherlands, which depended heavily on Ruhr coal).[35] In the course of the twentieth century, however, cartelization ambitions in the world coal industry were countered by the rise of additional coal exporters, distributed throughout the world. No "Coal OPEC" was ever formed and no single producing country or group of countries has been able to dictate world market prices. The economic risks linked to the international coal trade have thus been kept within limits. In fact, the perception of coal imports as more secure (in both physical and economic terms) than the other fossil fuels currently contributes greatly to the interests of many actors in retaining a large share of coal in their domestic energy systems.

Better known are the economic risks linked to global oil supply. While the 1973 Arab oil embargo remains an iconic supply disruption, the more lasting impact of the first oil crisis was the radical price hikes, enabled through OPEC cartelization in combination with an extremely tight world oil market, featuring minimal "shut-in" capacities. Ever since, fears of further price hikes and, more generally, turbulence in world energy prices, have been omnipresent. The 1980s and 1990s featured a temporary return to a market in which oil was sold and bought like a "normal" commodity,[36] but in the twenty-first century the fear of OPEC manipulating prices – now in cooperation with Russia – is again considerable. These economic fears constitute a major driving force in the transition from fossil fuels to renewables, and in particular, away from petrol- and diesel-fuelled vehicles to alternative transportation technologies.

Actors have often been able to counter economic risks by stimulating competition among exporters. In some situations, however, this may be easier said than done. Many Central and Eastern European countries, for example, are almost totally dependent on Russia for their natural gas supplies. While this has often been discussed in terms of the supply risks that such dependencies generate, the economic risks are even more pervasive. Countries such as Ukraine and the three Baltic nations have usually been charged much higher gas prices than would seem reasonable in view of their geographical proximity to Russia; Germany and France, which are located much further down the same pipelines, pay much lower prices. This asymmetry has often been interpreted in terms of Russia using natural gas exports as a foreign policy tool, but, above all, it reflects a lack of competition between gas from Russia and from elsewhere in Central and Eastern Europe. This makes the region vulnerable not only in terms of supply risks but also in terms of economic risks. Western European gas importers, by contrast, are able to exploit the healthy competition between Russia, Algeria, Norway and other suppliers, reducing economic risks of gas imports. It is only very recently that Central and Eastern European gas companies and state agencies have started to seriously embark on a diversification path. Concrete projects here include Lithuania's and Poland's new LNG import terminals and EU-supported investments in making existing East-West gas pipelines reversible.

Third, actors may oppose energy imports because of the *environmental risks* these present – at home and abroad. As noted above, the environmental risks linked to coal *mining* can be countered through substituting imports for domestic production. Yet the risks linked to the *combustion* of coal remain with the country in which the coal is actually burnt. Thus, not only coal-producing nations but also coal-importing nations have been plagued by the local, regional and global problems stemming from sulphur, nitrogen and carbon emissions. For some countries, this has made not only domestic coal production but also coal imports unacceptable or at least undesirable. Moreover, many environmental ministries, state agencies and non-governmental organizations (NGOs) have started to point to the moral responsibilities of advanced democracies to take into account the foreign environmental impact of energy imports. They refer to what we earlier called "entanglement" between, for example, local electricity systems and faraway coal and uranium mines. The latter become "shadow places"[37] linked to the domestic energy system, and the critics point to the need for the importing nations to make themselves aware of the far-reaching environmental destruction there – and, ideally, to phase out any imports that cannot be justified on environmental grounds.

Fourth and finally, importers face substantial *political risks* when engaging in the international energy trade. These are linked to a fear that energy imports will adversely affect the political freedom of action for a government, especially in the international arena. This will be discussed in further depth in Chapter 6. Suffice it here to note that the international energy trade has indeed been politically risky over the years, and there are many examples of countries whose governments have been forced to adjust their foreign policies due to their far-reaching energy dependence on other countries. Japan's radical remaking of its foreign policy *vis-à-vis* the Arab world in connection with the 1973 oil crisis is perhaps the best-known case, but it is also evident these days, for example, in Russian energy exports and how these delimit the EU's political room for manoeuvre.

Competing perceptions of opportunities and risks

Unsurprisingly, the interests of different actors often clash with each other in decision-making processes about engaging or not engaging in dependency-generating foreign relations. For example, in many cases, a high degree of energy dependence may be viewed as a problem from a foreign policy view, but as highly favourable from an environmental or economic point of view. This is the case, in particular, when domestic production of a certain energy source is expensive, environmentally hazardous or simply impossible in a physical sense. In other cases, foreign policy actors show enthusiasm for projects that do not seem economically or environmentally viable, while environmentally oriented actors may support projects that both foreign policy and business actors dismiss as overly risky or infeasible. Such clashes reflect the socially constructed nature of opportunities and risks.

The struggles between different interests in such cases are typically fierce and subject to intense public debate. The outcome reflects not only the power balance between different actors in a certain country but also the characteristics of *political cultures* and the *zeitgeist*. In Poland, for instance, foreign policy interests have long held the upper hand over economic and environmental interests. This has led the country to relinquish Soviet-designed nuclear power plants and minimize imports of Soviet natural gas, in favour of burning massive volumes of domestic hard coal and lignite. Energy dependence on the Soviet Union, so the Polish thesis, must be minimized at any price. As a result, Poland now has the fourth most carbon-intensive economy of all OECD countries.[38] In a similar way, Francoist propaganda in the late 1940s argued that "northern Spain must sacrifice the beauty of its landscapes for the nation";[39] from an energy independence point of view, it did not make sense to protest, as many locals had done for decades already, against local coal mining, however dirty and ugly it may be. Bulgaria, Estonia, Poland and Greece are other examples of countries that in a similar way deliberately sacrificed local environments for the sake of energy independence.[40] Another important – and tragic – case is Japan, which opted to build a breathtaking number of nuclear power plants on its territory to minimize energy independence – in spite of the widespread awareness of disaster risks stemming from earthquakes and tsunamis.

Elsewhere, actors opted to sacrifice independence for the sake of economy and environment. They saw opportunities rather than problems in becoming dependent on other countries. For example, Sweden phased out its uranium mining once an international market for uranium and enrichment services had been established, turning to imports. Imported uranium was cheaper, and imports from abroad also meant that the Swedes did not have to cope with any domestic environmental problems relating to uranium mining.[41] Fears of nuclear disasters also made many smaller nations turn to imports. Greece, for example, after years of debate, eventually chose not to build any nuclear facility on its territory; instead, the country coped with its energy shortages through scaled-up imports of electricity – much of it, ironically, taking the form of Bulgarian nuclear electricity.[42] Italy, in a similar fashion, abruptly shut down its nuclear reactors after the 1986 Chernobyl disaster, only to replace the energy they supplied with nuclear electricity imports from France. Armenia, after a devastating earthquake in December 1988, also closed down its one and only nuclear power plant. This had far-reaching consequences for Armenia's energy dependence in the turbulent post-Soviet era; in a globally unique response, one of the two decommissioned units went back online in November 1995.[43]

Coal mining in many European countries has also been gradually phased out in recent decades – not so much because the deposits have been physically depleted, but more because competition from other energy sources and from foreign coal supplies makes extraction economically unviable, and because coal mining has come to be regarded as environmentally hazardous. As a result, countries such as Germany and the Netherlands now import large volumes of coal from overseas, notably from the United States, Colombia and South Africa. However, resistance to the closure of coal mines has been strong from trade unions, not least in Britain. Hydropower projects, designed to come to grips with escalating foreign dependencies, have in many cases also been abandoned due to environmental concerns. The protests against new ambitious hydropower projects in Latvia and on the Slovak-Hungarian border in the late 1980s are probably the most telling examples of this.[44] Sweden's decision to "protect" four of its major rivers, whose exploitation could have significantly contributed to reducing the country's energy dependence in an economically efficient way, is another case in point.[45] Arguing along similar lines, Europe would potentially be able to resolve much of its present-day gas dependence on Russia by turning to intra-European shale gas exploration; however, most European governments have opted not to invest in this opportunity, citing environmental and other concerns. The implication is that the EU's gas dependence on Russia will persist for the foreseeable future.

Perceptions of opportunities and risks change over time. The changes are often slow, as in the case of solar energy and electrical vehicles slowly gaining ground as economically competitive, or the evidence for emissions-induced climate change gradually accumulating and becoming accepted. But sometimes perceptions change in sudden and radical ways in response to major *critical events*, in the form of dramatic accidents, economic crises, political revolutions and the like. Such events often redefine risk perceptions and the hierarchy of worst-case scenarios. In the years around 1970, for example, the US energy debate increasingly came to focus on various environmental risks. Then came the 1973 oil shock. Americans were shocked, and environmentalism suddenly receded into the background, while oil-supply risks came to be perceived as the most threatening of all risks. President Nixon promptly responded to the new situation. Shortly after the outbreak of the Yom Kippur War, he told the "alarmed and fearful nation" that he would take immediate action on several points. A massive "Project Independence" was launched, an undertaking in which environmental ambitions – and a lot of money – were sacrificed for the sake of supply security.[46] A few years later, when the second oil crisis hit, President Carter followed up Nixon's efforts by launching massive investments in expensive and environmentally dubious support for extracting oil shale resources in the Rocky Mountains and the production of "synthetic" liquid fuels through coal hydrogenation.[47]

Another case concerns the fluidity in the perceived relative importance of nuclear risks, climate risks and gas-supply risks in recent decades. Following the accidents at Three Mile Island and Chernobyl in 1979 and 1986, respectively, the nuclear energy industry faced a tough time. From the late 1990s, however, it saw a new upswing, sometimes formulated in terms of a proud "nuclear renaissance". The changing perception of nuclear energy as relatively risk-free was closely linked to growing concerns over climate change, skyrocketing oil prices and, in Europe, gas supply disruptions from Russia. The risks related to nuclear energy, its proponents argued, were far less serious than these problems. Then, in March 2011, Japan was shaken by the nuclear disaster at Fukushima. Risk perceptions suddenly changed again. In Germany, for example, the risk of gas supply disruptions from Russia and the climate risk stemming from the burning of coal started to be de-emphasized in relation to the perceived probability of nuclear accidents. In spite of massive investments in renewable energy production, Germany's coal importers saw their business grow as nuclear power plants were hurriedly shut down in the aftermath of Fukushima.[48]

Energy export dependence

The discussion so far has centred on the perspective of importing nations. These constitute a vast majority of the world's countries. But some countries – many of which belong to the developing world – are much more dependent on exports than imports of energy. The oil-exporting countries of the Middle East and North Africa are the most well-known among these, along with Russia, whose massive exports of a whole range of fuels – coal, oil, gas, uranium and even bioenergy – make it unique in the international energy arena. In the Western hemisphere, Mexico and Venezuela are the most export-dependent nations in oil, while Colombia has become heavily dependent on coal exports. In sub-Saharan Africa, Nigeria and Angola are the most export-dependent nations in oil, while Niger and Namibia rely on uranium sales and South Africa on coal shipments. Indonesia is another huge coal exporter, catering to the needs of the more advanced East and South-East Asian nations. Indonesia, along with neighbouring Malaysia, is also a world-leading exporter of biofuels. Among the most advanced Western nations, the most export-dependent ones are Norway, Canada and Australia. Further examples of export-dependent nations include the former Soviet republics of Kazakhstan (which supplies an astonishing 40% of the world's uranium needs), Turkmenistan, Uzbekistan and Azerbaijan. It may be added that North Africa and the Middle East supply not only oil but also natural gas; Qatar, in particular, is almost totally dependent on gas exports for the sustainability of its economy and society.

Energy exporters do not face any supply risks in the traditional sense. However, they face considerable economic risks. To start with, large revenues from fuel exports often generate macro-economic problems and imbalances. The "Dutch disease" is a case in point, signifying the risk that massive export income generates a trade surplus, which, in turn, increases the value of the local currency. Such a development, while making imports of various goods and services cheaper, weakens the competitiveness of domestic production on both domestic and foreign markets. Local non-energy industries may then be wiped out, and the dependence of the overall economy and society on energy exports increases further. The term "Dutch disease" was coined in 1977 to explain the decline of the Dutch manufacturing industry, which was interpreted as stemming from the rise of Dutch gas exports from its huge Groningen field.

A broader, widely used concept is that of the "resource curse", also known as the "paradox of plenty".[49] It stems from the empirical observation that resource-rich countries have often had immense problems turning fuel and other mineral wealth into sustainable economic dynamism and societal prosperity; it is as if resource wealth, paradoxically, were not such a blessing after all, but, on the contrary, a veritable curse. More recent research has tended to de-emphasize this phenomenon, noting that the "curse" is very selective: only some countries have, in fact, been affected negatively by their resource riches (whether this group constitutes a majority of all sizeable net exporters or not remains subject to dispute).[50] Of course, the term "curse" is also misplaced, because there is certainly no black magic involved here. Apart from the Dutch disease, what typically happens is that public and private actors fail to link success in fuel exports with innovation and entrepreneurship in energy-related equipment and services sectors; instead, the dynamism in these fields remains dominated by foreign players who act in isolation from the local economy. Norway is the most well-known counter-example here, with its sprawling and highly innovative cluster of companies, universities, research institutes and state agencies that support the oil and gas industry, generating value and expertise that is now much sought-after in many other parts of the world.

Norway has also earned a global reputation for its ability to cope successfully with the problem of massive resource "rents", so typical in the extractive industries. Rents in fuel extraction stem from the widely differing production costs at different extractive sites. In oil, for example, it costs Saudi Aramco no more than around $20 to produce a barrel of crude oil, whereas in other parts of the world, the production costs may be $60 or even $100. Under market economy conditions, the actual market price will depend on the production costs of the more expensive (the "marginal") producers. If the oil is sold on the spot market, every producer will get more or less the same price for its crude oil (although the exact price will depend on quality, volumes, etc.). As a result, the low-cost producers will earn profits that have no clear relation to their production costs. These exaggerated profits are termed "resource rents". The mere existence of such huge sums of money, especially in cases where the exporting country is poor, makes it tempting for local business and political elites to "capture" some of the rents, thus feeding corruption and abuses of the huge money flow. Whether or not this actually happens depends on the people involved, on the local traditions and political culture, and on the nature of formal rules, regulations and control mechanisms. Norwegian experts, having earned a reputation for their ability to avoid the worst side-effects of resource rents, now travel around the world offering their advice to governments in export-dependent nations.

Export earnings that are too *low* also cause problems, especially when the revenue streams change abruptly and unexpectedly. Exporting nations often use their export income for massive domestic investments external to the energy industry, notably various infrastructure projects (roads, railways, telecommunications, etc.) and a plethora of critical public services (schools, hospitals and so on). After some time (years, decades), the government has become so accustomed to using its fuel export income for public services that any significant drop in the revenue stream risks causing havoc to the state budget. In 2014, for example, a year of low oil prices, a large number of countries were highly troubled in this regard, having grown used to oil prices of $100 or more. State planners had counted on continued high prices. The exporters' state budgets required exceptionally high oil prices to be balanced. The "break-even" levels for some of them amounted to $140 (Iran), $121 (Venezuela, Algeria), $119 (Nigeria), $117 (Ecuador), $106 (Iraq), $98 (Angola) and $93 (Saudi Arabia).[51] But actual oil prices were much lower, having fallen from a record level of $160 in mid-2008 to a mere $51 by the end of 2014 for West Texas Intermediate (WTI), one of the main benchmarks in oil pricing. A year later, the price had fallen to $30. The result for the exporters was huge budget deficits, and some exporters were forced to turn to foreign lenders to cope with the situation. In the longer term, it seemed that they would have to cut government spending, with far-reaching effects for public services. Today we see the actual effects of this in the form of far-reaching public unrest and political instability in countries like Venezuela and Iran.

Renewable energy dependencies

How is the ongoing transition to renewables altering the "classical" patterns of energy dependence? Proponents of renewables often argue that renewables enable resource-poor countries to free themselves from their dependence on "problematic" exporters of fossil fuels and uranium. In other words, they believe that the transition to a less polluting and more environmentally friendly energy system goes hand in hand with greater energy security. The reality, however, is more complex.

First, some renewable energy sources generate dependencies that are not qualitatively different from fossil fuel dependencies. Bioenergy, in its various forms, is the most obvious example here. Just like oil, coal, gas and uranium, the world's bioenergy resources are unevenly distributed in geographical terms, stimulating international trade – and dependency relations – between the "haves" and "have-nots". To be sure, the resources are, to a certain extent, "mouldable": they can be increased through various investment and policy initiatives in a way that fossil fuels cannot. But because large-scale fuel crops cannot be grown everywhere, some regions have clear advantages over others. Brazil, for example, has long made use of its massive sugar cane plantations for bioethanol production. This activity, which was originally launched in response to the second oil crisis in 1979, enabled the country to emerge as a large-scale bioethanol producer – and exporter. At one point it seemed that Brazil was about to become a global ethanol superpower, spurring an energy-poor region such as the EU to forge a "strategic partnership" as well as an "energy partnership" with Brazil, both in 2007.[52] Subsequently, however, growing domestic needs in Brazil have led to a situation where the country is now a net importer of biofuels.[53]

The EU was self-sufficient in biofuels up to the early 2000s, after which imports started to grow at a rapid pace following a period of exponential growth in demand. In bioethanol, EU import dependence reached 33% of consumption in 2007 – with Brazil by far the most important supplier. EU biodiesel import dependence followed the same pattern, peaking in 2008 at 24% of consumption, although the countries of origin were different, with Argentina, the United States, Indonesia and Malaysia the main suppliers. As of 2007–2008 it seemed almost inevitable that EU bioethanol and biodiesel import dependence would continue to increase – and that the EU's oil import dependence, which biofuels helped to diminish, might thus be replaced by an equally troublesome biofuel import dependence. In what followed, however, EU biofuel imports were criticized on environmental grounds, and especially in relation to the looming food crisis in the developing world, in which skyrocketing food prices were partly attributed to edible crops and oils being diverted for biofuel production. Subsequent policy measures, including new import tariffs, then reduced the EU's import dependence, so that by 2017 only 10% of European fuel ethanol was imported. At the same time, overall biofuel demand levelled out.[54]

Secondly, the entanglements between renewables and a range of *critical minerals* generate new patterns of international dependencies. Rare earth elements (REEs), lithium, graphite, cobalt, platinum group metals, indium and gallium are just some of the minerals that renewable energy champions depend on. Here it is not so much energy production as such, but rather the system-building process that depends on foreign supplies. The availability – and market price – of the minerals influence the competitiveness of renewable energy sources *vis-à-vis* fossil fuels and nuclear energy. As a result, international conflicts over access to the desired minerals have become part and parcel of energy and geopolitics, and renewable energy policies in the Western world have become deeply entangled with large-scale mining activities in places such as Chile and Argentina (for lithium), the Democratic Republic of the Congo (for cobalt) and Inner Mongolia in China (for rare earths).

Minerals-related supply and economic risks are intimately linked with environmental risks. In the case of REE, the United States had been the leading global supplier up to the early 1990s. In 2002, however, the owners of the huge REE mine in Mountain Pass, California, opted to close down operations following depressed world prices and protests against the adverse environmental effects of REE mining in the region. The depressed prices were, in fact, largely caused by the aggressive expansion of Chinese

REE mining in the 1990s. By 2000, China totally dominated world REE production, and the rest of the world became heavily dependent on deliveries from the huge mines at Baotou in Inner Mongolia. Awareness of this dependence came to the fore in September 2010, when Western media reported that China had blocked rare earth shipments to Japan. Japan was totally dependent on Chinese REE supplies. The *Economist*, in this context, reported that "a disruption in [REE] supply would paralyze the Japanese economy as much as an oil embargo or food blockade".[55]

Third, the transition to renewables generates *new international electricity dependencies*. This is most obvious in regions such as Western and Central Europe, with its many smaller and mid-sized nations. These have always had a lot to gain from cross-border electricity connections, but the potential benefits become even more pronounced in the age of renewables. Most renewable energy production is intermittent, varying radically over time depending on weather and wind, whereas demand follows a much more even curve. There are two principle ways of coping with this imbalance: the first relies on energy storage solutions, and the second on load sharing in a geographically wider system. Both have attracted ample attention in recent years.

Efficient storage solutions could potentially pave the way for more decentralized electricity grids and thus reduce the need for long-distance – including international – transmission lines. The idea of load sharing, by contrast, focuses precisely on making maximum use of long-distance connections. The proponents argue that wind energy from the North and Baltic Seas, Scandinavian and Alpine hydropower, South European and North African solar energy and so on can complement each other in technically rational and economically beneficial ways. One part of this has to do with weather and climate conditions. By pooling wind energy from different regions, for example, the radical fluctuations in electricity production can be reduced at an aggregate system level. The same goes for solar energy, because "the sun always shines somewhere". Another part has to do with the economic potential of interconnecting different time zones. Here, East-West connections are the ones at focus. This has inspired China's State Grid, for example, to propose very long-distance transnational interconnections, through which Central Asia, Russia and Europe could potentially import Chinese solar energy when nightfall prevents the former regions from producing their own electricity from the sun – and vice versa.[56]

More realistic is probably the initiative, launched in 2009, to create a transnational offshore electricity grid in the North Sea to "maximize the efficient and economic use" of renewable energy sources in the region. Offshore wind energy and Norwegian hydropower are the main components here. The governments and transmission system operators of ten European countries are involved.

Moreover, a number of visions have been formulated regarding the potential for outright electricity trade (rather than the mutual sharing of available resources). The point of departure here is the observation that some nations will be in a position to produce vastly more renewable electricity than they could possibly consume domestically, and so might be interested in forging export deals. North Africa, in particular, has for the past two decades stimulated the imaginations of European electricity system-builders in this regard. Starting in 2007, especially, several extremely ambitious visions made headlines, notably the by now iconic Desertec project, championed by German private actors, and MEDGRID, advanced by the French government. Critics early on pointed to the risk of Europe becoming dependent on a number of Arab countries for their future electricity supply. The onset of the Arab Spring in 2011, in which several authoritarian regimes in the Mediterranean region were ousted, initially seemed to strengthen the prospects for far-reaching European-North African-Middle Eastern cooperation in the energy realm. In the next phase, however, conservative Islamic political forces seized power in several countries, while Islamist terrorist organizations gained ground and one of the bloodiest wars the region had ever seen erupted in Syria. As a result, the visions of large-scale electricity exports from North Africa and the Middle East to Europe now remain in the shadow of the renewable energy debate. However, some ongoing solar energy projects do look quite promising, especially in Morocco.[57]

From dependence to interdependence

As we have seen in this chapter, both importers and exporters are, as a rule, dependent on smooth, uninterrupted flows of fuel across borders – and so actors on both sides generally have an interest in maintaining the system's overall stability. Another way to put this is that importers and exporters are interdependent.

Energy interdependence becomes particularly pronounced in grid-based systems such as electricity and natural gas. This is because in such systems it is extremely difficult to alter the geography of cross-border flows – and of the corresponding revenue streams. For example, Europe may be highly dependent on gas imports from Russia, but the Russian state budget is, arguably, even more dependent on the export revenues that its gas sales to Europe generate. A total disruption of supplies would most probably send European gas prices skyrocketing, and, given the grid-based nature of the system, in some Eastern European localities a severe supply crisis might result because of the difficulties of bringing in large volumes of gas from elsewhere. But Gazprom would not be much better off, because the grid-based nature of the system prevents large volumes of gas from being shipped to alternative customers.[58] The company's finances – and with it the state's – would face severe strain, its stock would plummet, and maintenance works, along with investments in exploration, might have to be postponed. Russia might threaten Europe with an export embargo, but Europe might threaten Russia with an import embargo, whereby the latter might be further supported with a ban on exports of Western equipment, machinery and consultancy services. In the end, no one would really gain much, and that is precisely what characterizes any interdependent relation.

Most countries are more or less self-sufficient in the supply of electricity, but only under normal circumstances. As noted above, national and sub-national grids often rely strongly on each other for emergency capacities, and there are numerous examples of critical events where electricity companies in neighbouring countries have helped each other out. Over time, many grid operators, especially in Europe's interconnected power pools, have decommissioned domestic back-up generating capacities, as they fully trust the electricity companies in neighbouring states to come to the rescue in times of crisis. But the precondition for this is, of course, that everyone is willing to help everyone else.

In coal, oil, uranium and bioenergy, interdependency patterns are less clear-cut. Instead, what we see here is how long-term trends in the supply and demand relationship continuously shift the focus from import dependence to export dependence and back again. When world supplies tighten, prices go up and the leverage of exporters increases. This was what happened in the 1970s and again in the 2000s. When supplies increase and prices go down, as in the 1980s and in the 2010s, the situation changes; suddenly exporters are much more dependent on importers than the other way round, and concerns over security of supply give way to exporters' quest for security of demand. Such long-term swings are important to follow over time as they, more than anything else, create and destroy opportunities for actors to make use of energy as a foreign policy tool (see further Chapter 6). But the notion of such "leverage" should not be exaggerated; just as in the case of grid-based energy sources, both exporters and importers have an interest in maintaining the smooth, trouble-free flow of fuel.

As the complexity of global energy supply increases, so do global interdependencies. In the future, it might even become irrelevant to analyse energy in terms of "import dependence" and "export dependence", as these terms are increasingly overshadowed by ever more intricate syntheses of the two. There was a time when wood and then coal totally dominated world energy supply and Britain was the only large-scale fuel exporter. At that time, international dependencies were easy to analyse and understand. Not so in the twenty-first century. Nowadays the world still burns copious volumes of wood and coal, but in addition energy users harness, on shifting scales, the energies of oil and gas, of uranium and plutonium, of wind and sun, rivers and tides – and in each of them the geographical (inter)dependencies are extremely complex.

The interdependence synthesis is further strengthened by the whole range of entanglements discussed in Chapter 2, whereby the entanglement between energy supply and technology supply must be especially stressed. Nothing can stop the trend towards greater dependence, in global energy, on an ever-vaster range of advanced technologies. Most countries in the world will not have the capacity to produce more than a very small share of the technologies they will need – and so they will be almost totally dependent on foreign supplies in everything from nuclear transmutation services and electrical vehicle batteries to smart grid technologies and gas compressor equipment. And yet even a very small country might have a monopoly on one or the other critical component in, say, wave energy technology, and so might place itself in a position that cannot be ignored by others.

Exercises

- Look at the examples of paradoxical dependency relations discussed in this chapter. Can you find any additional cases?
- Pick a country and collect basic statistical information about its energy imports and exports. Then, deepen your analysis by scrutinizing the country's energy dependence from a *systems* perspective. How have the dependency patterns changed over time? Has the country ever been energy-independent?
- Can you identify any critical events that have altered the mainstream perceptions of opportunities and risks regarding your (or another) country's international energy relations?

Notes

1 Eurostat, "Energy Production and Imports".
2 See, for example, World Bank, "Net Energy Imports 1960–2015".
3 Hölsgens, "Resource Vulnerability and Energy Transitions in the Netherlands".
4 For example, Gustafson, *The Wheel of Fortune.*
5 See statistics provided by the WISE Uranium Project, www.wise-uranium.org/umkt.html.
6 Total crude oil production in 2016 amounted to 92.2 million barrels per day, 65.5 million of which were exported. See BP, *BP Statistical Review of World Energy 2017,* 14, 24.
7 Total natural gas production in 2016 amounted to 3,552 bcm. 737 bcm were exported by pipeline and 347 bcm in the form of LNG. See BP, *BP Statistical Review of World Energy 2017,* 28, 34.
8 BP, *BP Statistical Review of World Energy 2017,* 29, 34.
9 As of 2016, the US was still a net importer of gas, with 82 bcm coming in from Canada. BP, *BP Statistical Review of World Energy 2017,* 34.
10 IEA, *Coal Information.*
11 Rüdiger, "From Import Dependency to Self-Sufficiency in Denmark".
12 For example, "Egypt Faces Budget Crisis Because of Rising Oil Prices", *The Arab Weekly,* 4 February 2018.
13 "Latvia Plans to Boost Gas Storage Capacity to 2.8 bcm by 2025", *Reuters,* 3 October 2014.
14 See www.balticconnector.fi.
15 For example, "Unrest Grows amid Gas Rationing in Iran", *New York Times,* 29 June 2007.
16 Van der Vleuten and Lagendijk, "Transnational Infrastructure Vulnerability: The Historical Shaping of the 2006 European 'Blackout'".

17 Magnus von Bonsdorff, author interview, 10 November 2006.
18 The figures used here are taken from EURATOM Supply Agency, *Annual Report 2016*.
19 Yergin, *The Prize*, 691f.
20 "US Refineries Fast Running Out of Flexibility on Crude", *Reuters*, 9 October 2014; "Texas Flood: US Oil Exports Pour into Markets Worldwide", *Reuters*, 8 February 2018.
21 Högselius, "The Saudi Arabia of the Far East?", 414.
22 Högselius, *Red Gas*, 114f, 157, 165.
23 Vlček, "Critical Assessment of Diversification of Nuclear Fuel for the Operating VVER Reactors in the EU".
24 McPhee, "The Competition for the Ukrainian Nuclear Fuel Cycle".
25 Quoted in Yergin, *The Prize*, 704.
26 IEA, *Luxembourg*, 6.
27 For example, Högselius, "Spent Nuclear Fuel Policies in Historical Perspective".
28 Yergin, *The Prize*.
29 For example, Stent, *From Embargo to Ostpolitik*.
30 Izmestieva, "Integration of the European Coal Market and Russian Coal Imports".
31 "Uranbrytning blir förbjuden", *Svenska Dagbladet*, 17 May 2018.
32 Kaijser and Högselius, "Under the Damocles Sword".
33 Cordovil, "De-electrifying the History of Street Lighting"; Del Curto and Landi, "Gas-Light in Italy".
34 Shulman, *Coal and Empire*.
35 Hölsgens, "Resource Vulnerability and Energy Transitions in the Netherlands".
36 Yergin, *The Prize*, Chapter 35.
37 Cf. Plumwood, "Shadow Places and the Politics of Dwelling".
38 "Poland Needs a Strategy for Moving to a Lower-Emission Economy", OECD press-release, 23 April 2015.
39 Camprubi, "Whose Self-Sufficiency?"
40 Tchalakov et al., "Bulgarian Power Relations"; Holmberg, "Survival of the Unfit"; Cantoni, "Second Galicia?"; Arapostathis and Fotopolous, "Transnational Energy Flows".
41 Kaijser and Högselius, "Under the Damocles Sword".
42 Tympas et al., "Border-Crossing Electrons".
43 WNA, "Nuclear Energy in Armenia".
44 Fitzmaurice, *Damming the Danube*.
45 Kaijser and Högselius, "Under the Damocles Sword".
46 Yergin, *The Prize*, 617.
47 Ibid, 694.
48 Hence, the share of coal in German primary energy consumption grew from 23.2% in 2010 to 25.3% in 2011 – in spite of a parallel rapid growth of renewables. See *BP Statistical Review of World Energy 2012*, p. 41.
49 For example, Karl, "Paradox of Plenty".
50 For example, Mehlum et al., "Institutions and the Resource Curse".
51 *The Economist*, 28 January 2015.
52 "The European Union Deepens Energy Relations with Brazil", EU press-release, 5 July 2007.
53 Brazilian ethanol exports peaked in 2008.
54 Lamers et al., "International Bioenergy Trade".
55 Quoted in Vikström, "Specter of Scarcity", 2.
56 "China Looks to Export Surplus Energy to Germany", *Financial Times*, 30 March 2016.

57 See, for example, Tagliapietra, *Energy Relations in the Euro-Mediterranean.*
58 Recent initiatives for the construction of LNG facilities in both Russia (for diversifying exports) and Europe (for diversifying imports) must be seen in this light. Even so, Russia and Europe are bound to remain strongly interdependent in natural gas for the foreseeable future.

Further reading

Balmaceda, Margarita. *The Politics of Energy Dependency: Ukraine, Belarus, and Lithuania between Domestic Oligarchs and Russian Pressure, 1992–2012.* Toronto, ON: University of Toronto Press, 2013.
Lagendijk, Vincent. *Electrifying Europe: The Power of Europe in the Construction of Electricity Networks.* Amsterdam: Aksant, 2008.
Shulman, Peter. *Coal and Empire: The Birth of Energy Security in Industrial America.* Baltimore, MD: Johns Hopkins University Press, 2015.

Bibliography

Arapostathis, Stathis, and Yannis Fotopolous. "Transnational Energy Flows, Capacity Building and Greece's Quest for Energy Autarky, 1914–2010". *Energy Policy,* forthcoming.
BP. *BP Statistical Review of World Energy 2012.* BP, 2012.
BP. *BP Statistical Review of World Energy 2017.* BP, 2017.
Camprubi, Lino. "Whose Self-Sufficiency? Energy Dependency in Spain from 1939". *Energy Policy,* forthcoming.
Cantoni, Roberto. "Second Galicia? Poland's Shale Gas Rush through Historical Lenses". *Geological Society, London, Special Publications* 465 (14 May 2018): 201–217.
Cordovil, Bruno. "De-electrifying the History of Street Lighting: Energies in Use in Town and Country (Portugal, 1780s–1930s)". In *The Culture of Energy,* edited by Mogens Rüdiger, 30–81. Cambridge: Cambridge Scholars Publishing, 2008.
Del Curto, Davide, and Angelo Landi. "Gas-Light in Italy between 1700s & 1800s: A History of Lighting". In *The Culture of Energy,* edited by Mogens Rüdiger, 2–29. Cambridge: Cambridge Scholars Publishing, 2008.
EURATOM Supply Agency. *Annual Report 2016.* European Union, 2017.
Eurostat. "Energy Production and Imports". http://ec.europa.eu/eurostat/statistics-explained/index.php/Energy_production_and_imports (accessed 25 January 2018).
Fitzmaurice, John. *Damming the Danube: Gabcikovo/Nagymaros and Post-Communist Politics in Europe.* Boulder, CO: Westview Press, 1995.
Gustafson, Thane. *The Wheel of Fortune: The Battle for Oil and Power in Russia.* Cambridge, MA: Harvard University Press, 2012.
Högselius, Per. "Spent Nuclear Fuel Policies in Historical Perspective: An International Comparison," *Energy Policy* 37 (2009): 254–263.
Högselius, Per. *Red Gas: Russia and the Origins of European Energy Dependence.* Basingstoke and New York: Palgrave Macmillan, 2013.
Högselius, Per. "The Saudi Arabia of the Far East? China's Rise and Fall as an Oil Exporter". *The Extractive Industries and Society* 2 (2015): 411–418.

Holmberg, Rurik. "Survival of the Unfit: Path-Dependence and the Estonian Oil Shale Industry". PhD thesis, Linköping University, 2008.

Hölsgens, Rick. "Resource Vulnerability and Energy Transitions in the Netherlands since the Mid-nineteenth Century". *Energy Policy* (forthcoming).

IEA. *Luxembourg: Oil and Gas Security.* IEA, 2010.

IEA. *Coal Information: Overview.* IEA, 2017.

Izmestieva, Tamara. "Integration of the European Coal Market and Russian Coal Imports in the Late 19th and Early 20th Century." In *Integration of Commodity Markets in History,* edited by Clara Eugenia Núñez, 79–90. Proceedings of the Twelfth International Economic History Congress, Madrid, August 1998.

Kaijser, Arne, and Per Högselius. "Under the Damocles Sword: Managing Swedish Energy Dependence in the 20th century". *Energy Policy* (forthcoming).

Karl, Terry Lynn. *The Paradox of Plenty: Oil Booms and Petro-States.* Berkeley: University of California Press, 1997.

Lagendijk, Vincent, and Erik van der Vleuten. "Inventing Electrical Europe: Interdependencies, Borders, Vulnerabilities." In *The Making of Europe's Critical Infrastructures: Common Connections and Shared Vulnerabilities,* edited by Per Högselius, Anique Hommels, Arne Kaijser and Erik van der Vleuten, 62–104. Basingstoke and New York: Palgrave Macmillan, 2013.

Lamers, Patrick, Carlo Hamelinck, Martin Junger and André Faiij. "International Bioenergy Trade – A Review of Past Developments in the Liquid Biofuel Market". *Renewable and Sustainable Energy Reviews* 15 (2011): 2655–2676.

McPhee, Sarah L. "The Competition for the Ukrainian Nuclear Fuel Cycle: Rosatom, Westinghouse, and Implications for Nuclear Energy in the Near Abroad". MA thesis, University of Washington, 2015.

Mehlum, Halvor, Karl Moene and Ragnar Torvik. "Institutions and the Resource Curse". *The Economic Journal* 116 (January 2006): 1–20.

Plumwood, Val. "Shadow Places and the Politics of Dwelling". *Australian Humanities Review* 44 (2008): 139–150.

Rüdiger, Mogens. "From Import Dependency to Self-Sufficiency in Denmark, 1945–2000". *Energy Policy* (forthcoming).

Shulman, Peter. *Coal and Empire: The Birth of Energy Security in Industrial America.* Baltimore, MD: Johns Hopkins University Press, 2015.

Stent, Angela. *From Embargo to Ostpolitik: The Political Economy of West German–Soviet Relations, 1955–1980.* Cambridge: Cambridge University Press, 1981.

Tagliapietra, Simone. *Energy Relations in the Euro-Mediterranean: A Political Economy Perspective.* Basingstoke and New York: Palgrave Macmillan, 2017.

Tchalakov, Ivan, Tihomir Mitev and Ivaylo Hristov. "Bulgarian Power Relations: The Making of a Balkan Power Hub". In *The Making of Europe's Critical Infrastructures: Common Connections and Shared Vulnerabilities,* edited by Per Högselius, Anique Hommels, Arne Kaijser and Erik van der Vleuten, 131–156. Basingstoke and New York: Palgrave Macmillan, 2013.

Tympas, Aristotle, Stathis Arapostathis, Katerina Vlantoni and Yannis Garyfallos. "Border-Crossing Electrons: Critical Energy Flows to and from Greece". In: *The Making of Europe's Critical Infrastructures: Common Connections and Shared Vulnerabilities,* edited by Per Högselius, Anique Hommels, Arne Kaijser and Erik van der Vleuten, 157–186. Basingstoke and New York: Palgrave Macmillan.

Van der Vleuten, Erik, and Vincent Lagendijk. "Transnational Infrastructure Vulnerability: The Historical Shaping of the 2006 European 'Blackout'". *Energy Policy* 38 (2010): 2042–2052.

Vikström, Hanna. "The Specter of Scarcity: Metal Shortages in Historical Perspective, 1870–2015". PhD thesis, KTH Royal Institute of Technology, 2017.

WNA. "Nuclear Energy in Armenia". www.world-nuclear.org/information-library/country-profiles/countries-a-f/armenia.aspx (accessed 29 May 2018).

World Bank 2018. "Net Energy Imports 1960–2015". https://data.worldbank.org/indicator/EG.IMP.CONS.ZS (accessed 25 January 2018).

Yergin, Daniel. *The Prize: The Epic Quest for Oil, Money and Power*. London: Simon and Schuster, 1991.

5 Managing vulnerability in a geopolitical context

How can actors reduce their energy dependence on foreign nations, and how can they cope with their vulnerability to various threats in global energy? This chapter follows up on the preceding one by discussing the strategies and techniques that actors have historically deployed in response to growing energy dependence.

Actors in fuel-importing nations can pursue two overarching strategies. On the one hand, they can seek to *reduce* their reliance on foreign countries for their energy supply. On the other, they can try and *manage* their dependence. *Reducing* energy import dependence is essentially an *internal* challenge: it has to do with activities such as developing domestic energy sources and promoting domestic energy conservation. This indirectly allows the actors to reduce imports or at least prevent these from growing. Attempts to *manage* (rather than reduce) dependence comprise both *internal* and *external* activities.[1] Typical internal activities include creating strategic fuel reserves and developing rationing plans for emergency situations. External activities include efforts to diversify imports, various "energy diplomacy" initiatives and the creation of alliances and partnerships with other import-dependent nations. These variations on the overall theme of vulnerability management are discussed in further detail below.

Mobilizing domestic energy sources

The most obvious strategy that actors have pursued to *reduce* their reliance on foreign energy has been to substitute domestic energy sources for imported ones. Historically, this strategy can be traced back to the second half of the nineteenth century and the emerging coal dependencies of that era, especially in Europe. As industries and governments became aware that coal imports generated dependencies on other nations – especially on Britain, where the 1872–1873 "coal famine" gave rise to one of the first alarms among importers[2] – the "natural" response, in the eyes of many actors and analysts, was to look for possible domestic alternatives. Needless to say, coal-importing agencies did not necessarily agree, and they did their best to convince other actors that the benefits and opportunities linked to import dependence far outweighed the risks. But in most cases governments sided with the visionaries of autarky. These included not only hawkish foreign policy actors and industrialists with stakes in (potential) domestic fuel extraction, but also the national geological surveys that had been set up in most European countries by the end of the century.[3] Together with private explorative initiatives, these institutions led the way in the quest for greater national autonomy. Identifying domestic coal deposits was perceived as their most important task, but the search for alternative fuels also received attention. High hopes were placed, for example, in domestic peat resources as a means of increasing energy independence, as is clear from the energy histories of countries like Sweden and, later on, emerging nations such as Latvia and Lithuania.

The quest for domestic substitution of imported coal gained momentum in earnest following the trauma of coal shortages during World War I. The disruption of coal imports from Britain led to shutdowns of gas lighting in cities such as Lisbon and Milan, and in many places railway transports and industrial activities threatened to come to a standstill. As a consequence, exploration for domestic coal, peat, oil and even natural gas deposits – along with large-scale investments in hydropower – stood high on the agenda for interwar governments. It was in this context that many of the state-owned companies discussed in Chapter 3 were set up, ranging from electricity companies such as Vattenfall in Sweden and Imatran Voima in Finland to oil companies like CAMPSA in Spain and Agip in Italy. The domestic energy hunt intensified further after World War II.

The degree of success in domestic exploration efforts varied greatly both across countries and over time. In the immediate aftermath of World War II, Austria assumed a leading role in the European petroleum industry for a time, ensuring that country's energy independence in oil; yet, oil production soon peaked and then quickly declined. A similar development followed in the case of natural gas.[4] Germany, France and Italy also quickly scaled up oil and gas production, but likewise failed to sustain domestic growth in production, forcing these countries to turn to imports. Romania, for its part, managed to retain its role as an important oil and gas producer, in a way that was later to shape Romania's relative political independence vis-à-vis the Soviet Union.[5] Estonia, starting in the interwar era, set out to exploit its massive domestic oil-shale resources, at a pace that quickened markedly after the Soviet Union's 1944 annexation of the country.[6] The Soviet Union also stimulated local peat extraction in the Baltics, in Belarus and elsewhere. In East Germany, Czechoslovakia, Bulgaria, Yugoslavia, Albania and Greece, domestic lignite resources began to be exploited aggressively during the post-war decades, the poor quality of this fossil resource notwithstanding.[7] Francoist Spain, facing international isolation, also took on the challenge of identifying new coal deposits and exploiting them, in spite of the coal's low energy content.[8]

BOX 5.1 BULGARIA'S LIGNITE TRIUMPH

Bulgaria is one of many countries in the world that considers itself poor in domestic energy sources. However, Bulgaria does possess sizeable lignite resources. Although the lignite is of poor quality, enterprising Bulgarian scientists and engineers proposed that it might be possible to burn lignite on a commercial scale in an economically viable way. A first lignite-fuelled power plant, dubbed Komsomolska, was brought online in September 1960. It was built next to the country's largest lignite basin, Maritsa East, in the southeast of the country. The Komsomolska facility became an important addition to the Bulgarian energy system. It also became a key research and training centre for Bulgarian lignite extraction and burning. In the early 1960s, three Bulgarian scientists-engineers, including the country's later energy minister, Nikola Todoriev, enthusiastically set out to improve the original lignite-burning technology at Komsomolska. The point of departure was a rumour that engineers just across the border at Kozani in northern Greece, where lignite was also being extracted and burnt, had designed a new technology that eliminated the necessity of drying the coal and significantly increased the heat yield. The Kozani plant had been built by the West German electricity giant RWE.

> RWE used more modern and efficient fan mills for the coal than the Bulgarians did. To test the applicability of the technology to the less favourable properties of the Maritsa East lignite coal, in the mid-1960s, several trains loaded with Bulgarian lignite coal were sent through the Iron Curtain to Kozani for an experiment and burned by a team of Bulgarian, (West) German and Greek engineers.

The Bulgarians then set out to develop "a technology that replaced the existing hammer mills for coal with highly efficient fan mills, opening the door to the construction of larger power-generation units". As it turned out, these innovations "radically decreased costs and rendered Bulgarian lignite power plants among the most efficient at the time. Apart from making the coal-drying plant unnecessary, the energy needs of the thermal plant proper decreased from 14 per cent to 10 per cent".

This innovative project served Bulgaria well, as it counterbalanced its energy dependence on the Soviet Union. It made it economically feasible to radically scale up lignite mining in Bulgaria. There were only two problems. First, the power plants were extremely polluting, so that over time, they caused significant environmental damage throughout a large region.

For more than two decades, acid rain and heavy dust during windy periods became part of the lives of the local population. The situation did not improve until the 1990s, when most of the power-station chimneys were equipped with filters to capture the dust and the most dangerous compounds from the exhaust gases.

Second, the engineers worried that the eagerness with which they pursued their energy independence strategy might cause tensions in the country's relationship with the Soviet Union. When Todoriev was appointed minister of electrification in 1976, he expressed his fears that Bulgarian lignite, now being extracted and burned on a previously unseen scale, might aggravate the Kremlin. However, Bulgarian party leader Todor Zhivkov replied, "Your duty is to defend Bulgarian interests; there are other people in the Politburo who take care of the Bulgarian-Soviet friendship".[9]

Some countries were extremely lucky. In 1959, Shell and Esso, while looking for oil in the Netherlands, found one of the world's largest natural gas fields. Following complex negotiations and power struggles for control over this gas, the find seemed to ensure Dutch independence in gas. It also stimulated electricity and heat companies to substitute natural gas for other, imported energy sources. At the same time, the gas find opened up the opportunity for the Netherlands to become a powerful energy exporter with substantial potential leverage over other Western European nations, exercised through the tentacles of an emerging transnational pipeline network.[10]

At about the same time, explorations for oil and gas began in the North Sea, following a 1958 agreement on how to divide the continental shelf among the North Sea countries. Ten years later, the first major discoveries of oil and gas were made in the Norwegian sector, soon followed by discoveries in the British, Danish and Dutch parts of the shelf.[11] Norway soon emerged as one of the world's most important oil-producing nations. By 1980, Britain had become self-sufficient in oil, and Denmark followed suit in 1997.

In another part of the world, China, which earlier had been dependent on oil imports, successfully exploited the Soviet Union's ideologically motivated provision of technical assistance and expertise to the new People's Republic. This led to the discovery, in 1959, of one of the world's largest oil fields, Daqing in Manchuria. It radically changed communist China's international energy position. Allowing the country to free itself from imported hydrocarbons, Daqing guaranteed China's energy independence during the country's difficult period of international isolation in the 1960s and early 1970s. Daqing was later followed by additional large oil finds. Today, China still remains one of the world's largest oil producers.[12]

Other countries tried but failed to find significant fossil fuel deposits. Most of East Asia belongs to this category. In Europe, Sweden and Finland were among the unluckier: neither coal, nor oil, nor gas was found. Switzerland also failed to find oil on its territory, despite heavy investments in oil prospecting. However, all three nations attained a remarkable success in developing hydroelectricity – the "white coal".[13] Other smaller nations that invested heavily in hydropower were Austria, Yugoslavia, Spain, Latvia and, of course, Norway. In countries such as Greece, however, it soon turned out that the high hopes held for domestic hydropower had been exaggerated, especially in view of the radically growing domestic energy needs that quickly outpaced anything that national waterfalls could offer.[14]

Low oil prices during the 1950s and 1960s made many countries turn to imported oil as the very fundament of energy supply; in some cases, imported oil amounted to as much as 70–80% of a country's total energy needs (with Japan as one of the most dependent, at 78% as of 1973).[15] Oil dependence skyrocketed even in countries such as Germany, which sat on massive domestic reserves of hard coal and lignite. Oil became so cheap and so abundant that electricity companies opted to invest in oil-burning heat and power plants on a grand scale. The oil crises of the 1970s, however, generated a return to domestic coal as the backbone of the energy systems of many countries, and many oil-fuelled power plants were rebuilt so that they would run on domestic coal or (domestic or imported) gas rather than on imported oil. In the German case, this led to a halt in the planned long-term phase-out of coal.[16]

At the same time, the oil crises spurred a number of countries to intensify the development of another allegedly domestic energy source: nuclear energy. For the nuclear engineering community, the world turmoil in oil was a gift from above. Even though nuclear energy, as we have seen, should perhaps not be treated as a domestic energy source in the traditional sense, the nuclear community was very successful in framing it as such. The fundamental fact that none of the advanced nations in Europe or East Asia sourced their uranium domestically was downplayed. In this way, a country like France, having substituted massive volumes of nuclear electricity for the large amounts of oil that it previously imported, nowadays takes pride in its surprisingly low aggregate energy import dependence ratio (below 50%). If the nuclear industry's total dependence on foreign uranium supplies were taken into account, France's energy dependence would be almost 100%.

The dream of eradicating the need for energy imports through discoveries of vast new domestic energy sources is still very much alive today. In the case of oil, Brazil provides the most spectacular recent success story in countering energy dependence through domestic discoveries. The Brazilian finds were made in remote offshore fields in the South Atlantic Ocean. The oil is located at immense depths, and exploration and extraction are critically dependent on the most advanced recent technology. The 2010 Deepwater Horizon disaster in the Mexican Gulf hints at the environmental risks involved in undertakings of this kind, but the opportunities are also immense. Indeed, following its offshore bonanza, Brazil has become a net oil exporter and is well on its way to becoming one of the leading oil-producing countries in the world.

At the same time, the efforts to mobilize domestic energy sources have taken on new significance through the breakthrough of renewables and the parallel revolution in shale gas and oil. While the shale revolution is the one most clearly linked to energy geopolitics, the quest for renewables has been promoted by governments mainly on environmental (rather than security) grounds. The shale revolution, which has turned the United States from a net importer into a net exporter of natural gas, has so far been chiefly a North American phenomenon, although at the time of writing it appears to be spreading to a few other regions such as China. In Europe, shale oil and gas have captured the imaginations of state actors in countries like Ukraine and Poland, which look to shale as a potential escape from their energy dependence on Russia. For some time Poland appeared to be emerging as a European shale gas mecca, but subsequent prospecting and exploration have been disappointing and most actors have subsequently withdrawn. In Western Europe, meanwhile, the prospects for shale gas have been opposed on environmental grounds, occasionally leading to formal bans on prospecting and exploration.[17]

The efforts to reduce energy dependence by developing domestic energy sources thus continue in our own era. But the historical experience suggests that domestic energy is unlikely to solve the problem of energy dependence more than temporarily. This is most obvious in the case of fossil fuels, which are finite by nature and cannot be scaled up indefinitely. Sooner or later – usually in a matter of decades, but sometimes just a few years – fossil fuel producers face depletion of the available deposits, and after a brief golden age of relative energy independence a new, harsher era sets in. Ironically, the quest for domestic fossil fuels may even trigger new foreign dependencies. This was the case in Europe's attempts to use domestic natural gas resources to counter import dependence in coal and oil. In countries such as Italy, France, Austria and Germany, natural gas quickly became a popular fuel and demand grew rapidly. But the domestic producers soon faced exhaustion of the domestically available sources. Seeking to compensate for this, they looked to foreign suppliers to meet the expectations of domestic customers. Today all four countries are heavily dependent on imported natural gas.[18]

Internalizing non-extractive system activities

Actors in import-dependent nations may also seek to internalize non-extractive system activities, such as conversion and refinement, transportation, storage and waste management. The internalization of oil refining is a case in point. Historically, the international oil trade was dominated by refined products rather than crude oil, and the refineries were located chiefly in the major exporting nations. Over the years, as oil came to be perceived as a strategic commodity, oil-importing nations opted to construct their own refineries on domestic soil, so as to take control of this part of the system. The 1950s, 1960s and 1970s were the heydays of refinery construction in the Western world, and by 1980 the coasts of oil-poor Western Europe and non-Chinese East Asia were dotted by a string of refineries and petrochemical complexes. Not only the larger importing nations – like France, Germany, Britain and Japan – but most smaller nations, too, invested in domestic refineries roughly on a scale corresponding to their domestic needs. Many countries also invested in their own tanker fleets, so that they would not depend too much on the exporting nations and the large international oil companies for the transportation of their crude oil.[19]

In nuclear energy, importers have faced greater difficulties with internalizing refinery activities. Only the nuclear-weapons nations, along with a few other large nations such as Japan, Germany and the Netherlands, have found it technologically and economically feasible to construct their own uranium conversion and/or enrichment plants. All other countries with nuclear power plants remain dependent on foreign conversion and enrichment. France and Russia play particularly important roles here as suppliers of conversion and enrichment services to a range of smaller nuclearized nations. In the early days of nuclear energy, a lack of relevant knowledge and skills constituted the main obstacle to the construction of enrichment facilities; at that time nuclear enrichment was still a novel technology that most nations simply did not master. By the 1970s, the technology had matured and several large nations set out to profit from their own investments by exporting turnkey enrichment facilities to foreign countries. Since then, the main obstacle to internalizing nuclear enrichment has mainly been of a (geo)political nature, as the technology can also be used to produce highly enriched uranium for military purposes. Iran's controversial attempts to acquire nuclear enrichment capacity are a case in point here.

Internalizing refining and other non-extractive system activities may potentially even turn an import-dependent nation into an exporter (of refined products). For example, a country may build more domestic oil refineries on its territory than it actually needs, and take on a role as exporter of refined petroleum products. In Europe, Rotterdam and Antwerp early on became key "hubs" in the European oil system, importing vast amounts of crude oil. Some of this oil was simply reloaded onto smaller tankers and re-exported to other parts of Europe, but some was refined and exported as finished products. The two cities emerged as the most important European oil ports in a way that, from the perspective of Dutch and Belgian state and business actors, was seen to compensate for their far-reaching oil import dependence. On the other hand, Rotterdam's critical importance for European oil imports was a primary motive for the Arab nations to single out the Netherlands as a main target in the 1973 oil embargo.[20]

Reducing import dependence through technological development

Energy dependence can in some cases be reduced through domestic technological ingenuity. To follow up on the nuclear case, some countries early on sought to circumvent the problem of foreign enrichment by developing alternative types of nuclear reactors – notably heavy-water reactors – that could run on non-enriched uranium. Shifting their technological path, they thus hoped to avoid foreign dependencies in uranium enrichment. Canada soon emerged as the world leader here, while several European nations that originally hoped to develop heavy-water reactors subsequently abandoned these efforts. Today, countries such as India and South Korea remain committed to heavy-water reactor technology for parts of their nuclear programmes.

Another ambitious technology-oriented strategy for reducing energy import dependencies has been to develop synthetic fuels. Especially in coal-rich countries, this possibility has captured the imagination of both engineers and state actors. The main idea has been that import dependencies in oil and gas may be reduced through the hydrogenation of domestic coal. This is technically feasible but economically challenging, at least during periods of depressed world oil and gas prices. The radical efforts of Nazi Germany to substitute synthetic coal oils for imported crudes during World War II are the best-known example. Following the energy crises of the 1970s, the Carter administration in the US launched another ambitious "synfuels" investment programme.[21] In our own era, the idea of coal hydrogenation is still alive and kicking, although its popularity depends on high market prices for oil and gas. In addition, forest-rich countries are increasingly taking an interest in synthetic fuels, eyeing the possibility of turning wood into oil and gas. This is very much what "second-generation" biofuels are about in forest-rich countries like Sweden and Finland. Attempts to substitute wood-based jet fuels for traditional aviation kerosene, in particular, keep many a scientist and engineer busy these days, the motivation being not only environmental but also geopolitical. It may be added that wood already came to the rescue in some countries during World War II, at which time numerous vehicles in Europe's oil-poor countries, facing scarcity of imported oil, were equipped with specially designed wood gasifiers. In this way Sweden, for example, was able to keep its vehicle fleet rolling when the country's oil imports were disrupted.[22]

BOX 5.2 HITLER'S HYDROGENATION PROGRAMME

Germany has historically been a leader in modern chemical engineering. It also holds some of Europe's largest coal reserves. Combining these two advantages, German scientists and engineers emerged as pioneers in the production of coal-based synthetic fuels. As early as 1913, Friedrich Bergius patented the process that still bears his name, which enables the direct conversion of coal to synthetic liquid fuels. Even better known is the process for indirect coal conversion developed by Franz Fischer and Hans Tropsch in the 1920s. In the Fischer-Tropsch process, coal is first gasified and then transformed into various refined oil products. During World War II, with Germany facing cut-offs of crude oil supplies from overseas, Hitler's engineers scaled up early "synfuels" production efforts to unprecedented levels, seeking to meet wartime oil requirements of everything from diesel to aviation kerosene. Both the Bergius and the Fischer-Tropsch processes were put to work in this context, with large plants erected at Leuna, Pölitz, Ludwigshafen and elsewhere. Some were built underground to protect them from Allied bombings. By 1944, the Reich was producing around 124,000 barrels of synthetic oils per day. Yet the requirements were far larger than that, and Hitler's failure to access sufficient volumes of hydrocarbons is often cited as one of the reasons behind Germany's defeat in World War II.[23]

Figure 5.1 Bombing of Auschwitz, August 1944. The infamous Auschwitz-Birkenau concentration camp was one of many sites in Germany where factories making synthetic fuel were located. The military importance of synthetic, coal-based fuels turned the factories into targets for Allied bombing raids. Photo: AKG/TT.

Investments in domestic technological development can also help countries to reduce their dependence on foreign supplies of equipment, machinery, components and services of various kinds. The large industrialized countries clearly have the greatest chances of achieving independence in this way, although in today's globalized world of technology even the great powers depend to a great or lesser extent on foreign supplies of one item or the other. As for the world's smaller nations, as a rule these are heavily dependent on foreign suppliers for everything from nuclear reactors and oil drilling equipment to liquefied natural gas (LNG) plants and high-voltage transmission lines. Eastern Europe's heavy dependence on Soviet nuclear technology is a case in point, whereas Greece, Denmark and Sweden found that they had to rely on American and other foreign technology when constructing domestic oil refineries. In the same vein, when the Nordic countries embarked upon ambitious domestic oil exploration projects in the 1960s, they had little choice but to seek cooperation with the large international oil majors, becoming dependent on them for access to key equipment and know-how.[24] Spain, meanwhile, drew on first German, and later American technology when developing facilities for synthetic fuel production (on the basis of bituminous shale).[25]

However, there are a number of encouraging cases of smaller nations actually having managed to avoid the problem of technological dependencies on foreign nations. The most striking example is perhaps Sweden's all-domestic design of nuclear power plants, which became a source of much technological pride in the 1960s.[26] Both Sweden and Switzerland, their smallness notwithstanding, also became world leaders in high-voltage transmission technology and in the construction of turbines for hydroelectric power plants.[27] In communist Eastern Europe, Czechoslovakia retained a dominant technological position in key segments of the equipment industry, not least in the nuclear field. More recently, Denmark's and Spain's rise to dominance in wind turbine production exemplifies how smaller countries may play important technological roles in an age of renewables.[28]

Energy conservation for security purposes

A country's import dependence can be reduced substantially through *energy conservation*, that is, through reducing overall energy consumption and increasing energy efficiency. The main goal of energy conservation has usually been to minimize economic risks in the face of looming fuel imports, and to counter environmental problems related to the combustion of fossil fuels. Fuel exporters have traditionally been less interested in energy conservation than the importers. More often than not they have, in fact, encouraged wasteful energy use through generous state subsidies and politically controlled prices. The need to remove such subsidies is now increasingly being recognized, but doing so is bound to cause protests and dissatisfaction among industrial users and the general public. For this reason, many governments are hesitant to take action. They are most motivated in countries where rapidly growing domestic consumption threatens to put an end to energy exports. Today, even the large oil producers in the Middle East take interest in energy conservation, as an ever-larger share of domestic production is used internally, thus reducing the amount of fuel available for foreign sales.

As for the importing countries, Germany and its response to the oil crises of the 1970s illustrates what state actors can actually do to promote energy conservation. There was little interest in the issue before the late 1970s. In 1974, the government's energy experts still took for granted that energy consumption was bound to grow roughly in proportion to GDP, and none of the main political parties wished to see a decrease in GDP. Hence, it did not seem to make macro-economic sense to promote energy saving. In contrast to many other import-dependent countries, however, Germany did not take any measures to protect domestic energy users from the radical oil price hikes of 1973/1974. This was in itself an important (non)move that had a clear impact in terms of reduced overall energy consumption. Then, in 1976, the government took further action. Overall views of energy in society were undergoing far-reaching change. Energy conservation not only was becoming widely accepted as something desirable but was now even emerging as a top priority in energy policy. To begin with, the government passed an energy savings law that allowed it to establish standards for thermal insulation in buildings. This was followed up two years later by a House Modernization Act, which required house builders and renovators to improve insulation and heating systems in all buildings. At about the same time, the government radically raised the tax on heating oil and on petrol and diesel. It also encouraged the recovery of waste heat from electric power plants and combined heat and power plants, notably through tax deductions for such investments. In addition, Bonn entered into "voluntary" agreements with the domestic vehicle industry to improve the fuel standards of trucks and cars, threatening with binding legislation if manufacturers refused to cooperate.[29]

Strategic stockpiling and system flexibility

Shifting our focus from efforts to *reduce* foreign energy dependencies towards attempts to *cope* with dependence, strategic stockpiling comes to the fore as a first strategy. Here it should first of all be emphasized that stockpiles and storage facilities of different kinds are of considerable importance, not only in the context of emergency situations but also, and above all, for the everyday functioning of energy systems. The most iconic image of a storage facility is probably that of an oil cistern in a major seaport or a coal pile next to a large power plant; these are necessary to enable the everyday flow of fuel from geological deposits to power plants, filling stations and end users. *Strategic* stockpiles are different, being much larger and not necessarily stored visibly above ground. The US Strategic Petroleum Reserve, to take the most extreme case, has an astonishing capacity of 727 million barrels of crude oil. It is located in the heartlands of the US oil industry in Louisiana and Texas.[30] The Chinese counterpart has a capacity to hold up to around 300–400 million barrels. The International Energy Agency (IEA) mandates its member states – except those who are net exporters – to keep strategic petroleum reserves equivalent to at least 90 days of the previous year's net oil imports.

Early strategic reserves of energy were established by military actors, but with the rise of coal and, especially, oil as fuels critical for entire economies and societies, the stockpiles were radically scaled-up to cover emergency civilian needs as well. Unsurprisingly, the most import-dependent nations were the first to take such action. France was the first major country to establish a national strategic petroleum reserve, with a 1958 government decree requiring all oil importers to keep a crude oil reserve equivalent to at least three months of domestic sales.[31] In 1965, the West German government followed suit by requiring "refiners of crude oil and importers of refined petroleum products to maintain stocks of 65 and 45 days, respectively".[32] In December 1968, following the Six-Day War in 1967, the European Economic Community (EEC, the European Union's predecessor) issued a directive requesting all member states to "maintain, at all times … their stocks of petroleum products at a level corresponding … to at least 65 days' average daily internal consumption in the preceding calendar year".[33] The Japanese oil industry began a stockpiling programme in 1972, initially without formal government involvement.[34] The United States, by contrast, being much richer in domestic crude oil, established its Strategic Petroleum Reserve only in 1975, in response to the 1973/1974 oil crisis. Before that it relied on its "shut-in" oil production capacities.

Coal has also been subject to strategic stockpiling. In the post-war era, for example, the German government had a hard coal reserve of 10 million tons, corresponding to about an eighth of annual production.[35]

Strategic stockpiling of coal and oil is fairly straightforward. The same cannot be said for *natural gas*. When several Western European countries embarked on large-scale gas imports starting in the mid-1960s, the issue of strategic storage for emergency situations was soon raised. Since keeping gas in artificially constructed facilities – such as caverns or LNG storage facilities – is extremely expensive, the main solution that initially emerged centred on the "shut-in" principle that had proven its value in the international oil trade (see Chapter 2). Austria, for example, facing frequent irregularities and unintentional disruptions in its early gas imports from the Soviet Union, was able to successfully cope with these problems by temporarily increasing its own gas production. At that time, most European countries possessed such shut-in capacities in gas that could be called upon if needed.

Later on, with domestic reserves shrinking, the technique was adapted in such a way that imported gas could be physically fed into partly depleted domestic gas fields. In that way, the shut-in capacities could be artificially increased. A technically more challenging version of this storage method was to identify bedrock capable of absorbing large gas volumes. In the 1980s, for example, West Berlin found that its subsoil might be mobilized for this purpose. This became the cornerstone in the city's strategy to switch from coal to gas as its main energy source. The background was that the gas would be supplied exclusively from the Soviet Union. Fears mounted that such dependence might strangle West Berlin in the case of a geopolitically motivated supply disruption. Gas companies and state actors thought that the envisioned gas storage facility, with a capacity equivalent to a full year's fuel needs, would counter this nightmare.[36]

While rarely acknowledged in the political debate, storage is also a defining element of vulnerability management in *nuclear energy*. A key advantage of uranium-based fuels is that they are very compact. They require only very limited space when kept in storage. Nuclear power plants take advantage of this. In the typical case, a full batch of fresh nuclear fuel is kept right next to every reactor, and this batch guarantees the reactor's continued full-scale operation for a year or more. Hence, even in the case of a total disruption in the international nuclear fuel trade, the production of nuclear energy can go on for a very long time – much longer than the 90 days nowadays used as the benchmark for strategic storage of oil. As we saw in Chapter 4, the supply security is perceived to be so high that nuclear energy is often even thought of as a domestic energy source – even though the fuel, as is generally the case, is actually imported!

Strategic stockpiling can be regarded as part of a wider strategy aimed at creating flexibility in the domestic energy system. On the supply side, another measure that can be taken here is to build power plants that, without great additional costs, can switch from oil to gas, from coal to oil or the like. Construction of such power plants became very popular after the 1973/1974 oil crisis. Europe built no less than 43 GW of dual- or multi-fuel fired thermal power plants between 1974 and 1990, out of a 52-GW net increase in overall capacity. In most such flexible plants, switching to an alternative fuel can occur within a day in the case of a supply crisis.[37] On a more refined level, actors have also tried to make sure that power plants, whether they run on coal, oil, gas, uranium or renewable fuels, are not locked into supplies from a specific coal mine or oil field. Denmark, when seeking diversification away from British coal in the post-war decades, thus made sure that coal supplies from Poland would be interchangeable with British and American supplies.[38] Ensuring interchangeability of supplies has at times been costly, however; this was the case, for example, when Bavaria, in the years around 1970, built up a gas supply system based on a combination of supplies from the Soviet Union and the Netherlands; for substitution to be possible in case of a supply disruption from the Soviet Union, expensive conversion equipment had to be installed, because Dutch gas had a much lower calorific value.[39] The often prohibitive costs of such arrangements have, in practice, often led to decisions to accept a greater degree of dependence.

A further flexibility-enhancing strategy is for energy companies to conclude contracts with "interruptible" customers. These are usually customers in the industrial realm that are able to manage a major supply disruption and are prepared to accept a supply risk in return for lower gas or electricity prices. For example, as of the early 2000s, the German gas industry had interruptible contracts covering about 10–20% of all gas consumption.[40] This is an example of vulnerability management being approached from a demand rather than supply side. Gas and electricity companies have often found such arrangements economically advantageous as an alternative – or supplement – to expensive storage facilities and the like. When adopted on a large scale, the principle of interruptible customers forms a cornerstone of system-wide stand-by rationing and allocation plans, to be executed in the case of a major international supply crisis.

Strengthening domestic system governance

Another method to cope with vulnerability has been to seek *state control of energy imports*. Such initiatives were first discussed in earnest in the context of World War I and its aftermath. In continental Europe, the Spanish government became the first, in 1927, to declare a state monopoly over oil imports and oil refining.[41] Soon afterwards France and Germany followed suit by taking radical measures to strengthen government control over oil flows, causing much headache for the international oil companies.[42] State involvement gained further momentum after World War II, although the state's quest for control was often controversial. In Sweden, for example, where the state had taken control of wartime energy imports, post-war efforts to form a national oil-importing company eventually failed. The responsible Swedish actors feared that nationalizing oil imports might lead to conflicts with the large international oil companies, on which the country would anyway remain dependent. The conclusion was that the country's oil import security had nothing to gain from the establishment of a national oil agency.[43] The Danes, too, while not contemplating the creation of a state oil company, feared possible countermeasures from the international majors in the context of Denmark's state-led attempts to import large volumes of Soviet oil in the 1950s. In natural gas, Denmark did create a wholly state-owned company, DONG, the initial purpose of which was precisely to coordinate gas imports and counter vulnerability to foreign manipulations. Only a state company, it was believed, would be able to negotiate in a fruitful way with powerful foreign suppliers.[44]

Apart from directly controlling fuel imports, governments have also sought to forge cooperation among the nation's private energy actors in international issues. In Sweden, for example, the state urged the country's nuclear power plant operators to merge their nuclear fuel import activities, and a special company – Swedish Nuclear Fuel Supplies Ltd. (SKBF) – was set up for this purpose. The underlying idea was very similar to that of a state-owned fuel-importing agency: to strengthen the country's bargaining power in negotiations with foreign suppliers, both in contractual negotiations and in potential disputes over supply problems.[45] Many smaller countries have found it necessary to follow such a path, sensing their relative weakness in the international arena. But the strategy may be controversial, as it sometimes contradicts other state interests, such as the desire to maintain a competitive domestic energy market. From this point of view, it is not always desirable that the energy companies in a country cooperate closely.

The quest for diversification

Import-dependent nations have often sought to manage their energy dependence by *diversifying their energy imports*, in terms of both energy sources and supplier countries. The vision has been to attain a sound energy mix based on not only one, but several different fuels, each of which should ideally be sourced from a different set of countries. The idea is, of course, that should supply from a particular supplier be disrupted, the consequences will not be as severe and that the importer may even be able to compensate for the loss by temporarily bringing in imports from elsewhere. In addition, diversification has generally been seen as necessary when seeking to hedge economic risks in the form of potential monopoly or oligopoly abuse by dominant suppliers.

Smaller countries have generally found it more difficult to diversify their energy imports than large countries. A country such as Luxembourg, to take an extreme case, simply cannot afford to spread its risks by concluding deals with several different oil or coal suppliers. As a result, today Luxembourg, which does not have any domestic oil refinery, is almost totally dependent on Belgium for its access to oil products and almost totally dependent on South Africa for its coal supplies. Still, most countries, even the smaller ones, have historically been very active in seeking diversification, and have often been remarkably successful in this respect. Back in the nineteenth century, the Netherlands, for example, had already managed to counter Britain's dominance on the Dutch coal market by bringing in Belgian and German supplies,[46] while Denmark, in the post-war years, found that it could use Polish coal imports to put pressure on Britain to come up with attractive coal offers.[47] It was in coal that companies and governments in many energy-poor countries learnt to use the diversification of suppliers as a strategic tool in coping with energy dependence, and that experience became a useful stepping stone for subsequent diversification efforts in oil, gas and uranium.

Diversification has been more difficult to achieve in natural gas, mainly because of the predominance of pipe-based transports of this fuel. To this day, Sweden, for example, remains totally dependent on a single pipeline that brings in gas from Denmark; a rupture here would cause havoc for much of southern and western Sweden.[48] Finland, similarly, is supplied through a single pipeline from Russia. Spain, for its part, has found it remarkably difficult to diversify its gas imports away from Algerian supplies, although in recent years, additional piped gas from Norway along with large-scale LNG supplies from Nigeria and Qatar have yielded greater success.[49] The most striking cases of non-diversification are found in Central and Eastern Europe, where the historical legacy of Soviet gas in combination with lack of capital for new pipeline projects has resulted in a situation where Russia continues to totally dominate the market. Several recent initiatives, however, are arguably changing the picture. To these belong Lithuania's and Poland's investments in LNG terminals and various projects that, with ample support from the EU, aim to make Central and Eastern European pipelines reversible.[50] Bulgaria hopes to diversify its gas supplies by accessing North African gas, to be piped through Greece.[51] Armenia is one of the few ex-Soviet republics to actually have diversified away from Russian supplies, by bringing in Iranian gas, but since Gazprom is the owner of the Iran-Armenia pipeline, the actual value of these additional supplies is subject to debate.

The stories of Germany, France, Belgium, Switzerland and Italy offer a striking contrast to that of the above countries in Europe's periphery. They have been exceptionally successful in diversifying their gas imports. Here, the logic of diversification can even be regarded as one of the main forces driving the spectacular scale-up of Western European gas imports from the late 1960s up to the present day. A growing share of European gas supplies has come from what state actors have regarded as unreliable exporting nations, such as Russia and Algeria. The political argument has been that imports from these countries are still acceptable – as long as none of them provides more than a third or so of total supplies. (The exact percentage that is regarded as "acceptable" has varied while also being subject to intense political disputes.) In that way, it has been possible to radically increase European imports of both Russian and Algerian supplies, because supplies from allegedly secure sources – such as Norway – have also been scaled up. The absolute volumes of gas coming in from Russia and Algeria are now vastly greater than a few decades ago, but their *relative* importance in the gas mix has been kept within limits. The availability of intra-EU gas from the Netherlands has further strengthened the legitimacy of scaled-up imports from "unreliable" sources. France has been the most successful Western European gas importer in terms of diversification: it produces almost no domestic gas, but since no single exporter (with the exception of Norway) controls more than a fifth of the French gas market, the country's vulnerability in the geopolitical context is perceived to be very low – especially since natural gas in no way dominates the country's overall energy supply.[52] But Europe is probably the only region where such successes are possible. South America, for example, has seen attempts to emulate the internationally integrated European gas grid, but the development has not been terribly successful. A major reason is that there are simply not that many potential large-scale exporters around (see further Chapter 7).

The diversification strategy was less prominent in the massive up-scaling of global oil consumption in the post-war era. Many importers became highly dependent on just one or a few exporting nations, or on the Middle East as a region. The situation changed in earnest only after the oil crises of the 1970s. The German government, for example, while revising its energy policies in 1974, called for limiting Germany's imports from any single country to a maximum of 15%.[53] Still today, however, many import-dependent nations remain heavily dependent on one or a few countries in the Middle East for its oil supplies. China is the country that has most aggressively sought to diversify away from that region. Chinese state-owned energy companies have done so primarily by turning to imports of African and Central Asian oil and, as part of this effort, investing in the oil industries of these countries to stimulate a further surge in available supplies.[54]

Military action, foreign investment and energy diplomacy

The discussion so far has focused on *internal* attempts to cope with energy dependence. We now come to *external* approaches – that is, attempts to come to grips with vulnerability through action abroad. The connection between vulnerability management and geopolitics, which often remains indirect and invisible in the internal approaches, here becomes much more explicit. External approaches comprise a colourful palette of strategies and initiatives. At one extreme, external action may take the form of military attacks and other forms of physical violence and aggression. At the other, it may refer to peaceful foreign investment activities, energy diplomacy initiatives and various cooperative endeavours.

The most radical external approach is to mobilize armed forces to enforce access to foreign fuel supplies, refinery operations and transportation routes. The quest for energy and other natural resources has long been part and parcel of ruthless empire-building and colonialist expansion, from resource-poor Japan's conquest of Manchuria and Southeast Asia in the 1930s to Saddam Hussein's vision of a Greater Iraq in the 1980s and 1990s.[55] Energy has also been a salient feature in numerous international wars. In the two world wars, resource-rich regions such as Romania and the Caucasus early on became subject to violent military attacks and conquest attempts. Other well-known examples include the Iran-Iraq War (1980–1988) and the two Gulf wars (1990–1991 and 2003).

More commonly, empires and great powers have called on their militaries to "protect" foreign extractive activities and transportation routes, especially when these have been deemed to be of "strategic" importance. Britain's Royal Navy played an important role in protecting Middle Eastern oil operations from their launch in the early twentieth century until its withdrawal from the region in 1971. The United States then largely took over that role. Military escorts of oil tankers became a common sight on the high seas during World War II. American and French military vessels offered similar services – although this was highly controversial both domestically and internationally – during the Iran-Iraq War.[56] Further East, China is now considering the possibility of having the People's Liberation Army Navy (PLAN) escort its growing fleet of Chinese-flagged oil tankers.[57] Meanwhile, the United States has provided military training for a new task force in Georgia, the purpose of which is to protect the new Baku-Tbilisi-Ceyhan oil pipeline.[58]

Sometimes, outright attacks on foreign nations have been deemed "necessary" to maintain a normal flow of fuel from exporting to importing nations. In the 1956–1957 Suez Crisis, Israel, Britain and France attacked Egypt in an – eventually failed – attempt to regain Western control of the Suez Canal, a critical waterway for European oil imports. At that time, the United States refused to support the European action. In 1980, however, Washington officially announced what came to be known as the Carter Doctrine: that "the uninterrupted export of Persian Gulf oil" was essential to the US economy and that any move by a hostile power to disrupt that flow would be "regarded as an assault on the vital interests of the United States of America". As such, it would be "repelled by any means necessary, including military force". The policy was triggered by the Soviet assault on Afghanistan in 1979, which seemed to bring the communist superpower's armed forces threateningly close to Middle Eastern oil operations.[59] Similarly, France's military intervention in Mali is one of many international conflicts in our own era that cannot be understood without taking into account the need to protect global energy flows; at stake in this case are the huge uranium flows from the Saharan desert, of critical importance to French nuclear fuel supply security.[60] Many other "peace-keeping" missions are related in one way or another to importing nations' quest for stability in the global energy system. However, military interventions in fuel-rich countries by Western democracies have always been extremely controversial – both in the "target" countries and in domestic public opinion.

A slightly softer, but still radical and not necessarily peaceful approach has been to support or even orchestrate political coups in exporting countries, with the underlying idea that a more "friendly" government will help to ensure stability in fuel exports. The US-led coup in Iran in 1953 is a prominent example. Shortly before, Iranian Prime Minister Mohammad Mosaddegh had taken the radical decision to nationalize the country's oil industry and expropriate the Western oil companies that had been responsible for its creation. Both the British and the American governments concluded that the nationalization was inimical to their national security priorities. Mosaddegh was then overthrown "in a coup stage-managed and financed by their intelligence services".[61] However, energy is rarely the only item at stake in such coups. Indeed, actions like the one in Iran must be seen in a wider geopolitical context, as they typically involve broad foreign policy interests such as keeping rivalling great powers out of the country in question. This was especially common during the Cold War, when the United States and the Soviet Union fought for dominance in numerous oil-producing countries. It continues today in regions like Central Asia.

Companies and governments have also sought to gain control over foreign fuel riches in less violent ways, of course. Most of the oil concessions that Western nations historically secured in the Middle East, North Africa, Latin America and elsewhere were the outcome of more or less peaceful negotiations. To be sure, the talks that paved the way for such agreements were not always carried out in a friendly atmosphere, and there was sometimes a latent threat of violence in the air, especially when people and businesses linked to the great powers were involved. But more often than not, both parties – typically a private oil company and the monarch of an oil-rich territory – felt that they had something to gain from cooperation. The historical oil concessions that were born in this way allowed a number of highly import-dependent nations – notably Britain, France and Italy, but also Germany and Japan – to take control of some of the foreign oil they depended on. Such concession agreements later came under heavy political and popular attack in the exporting nations. They were now deemed to be unfair, with actors in the exporting nations concluding that they were being exploited by Western imperialism and capitalism. This shift coincided with a new era of decolonization, new nationalist movements and general anti-Western sentiment in the Middle East and elsewhere. This eventually led to nationalizations of Western – and Japanese – oil operations in many producing countries and, more generally, to a phase-out of the historical concession system. The 1973 Arab Oil Embargo marked the climax of this development. Yet the Western – and increasingly Asian – quest for foreign upstream investments as a way of coping with energy dependence has remained an attractive strategy for large energy companies, private and state-owned alike. It is merely taking a new form. The massive investments by Chinese state-owned companies in African fuel extraction, refining and transportation are a case in point. Japan's now failed attempt to gain ownership of some of Russia's Siberian and Far Eastern fuel riches is another. The French nuclear industry's far-reaching control of uranium mining in Niger is a third.

Action through upstream investments in the producing nations as a vulnerability-management strategy has usually been associated with great power geopolitics. But the smaller import-dependent countries have typically sought to replicate the great powers' external actions. As a matter of fact, they have been more active in this regard than is usually assumed. One of the earliest examples is Sweden's establishment – through a joint public-private initiative – of a coal mine in Spitsbergen in the early twentieth century, a main motive being to counter Sweden's problematic dependence on British coal.[62] More recently, Swedish state and business actors have acquired sizeable German lignite mines and important stakes in Sudanese oil fields. The Danish and Spanish governments similarly sought to emulate the aggressive activities of the large imperial powers by exploring their own colonial possessions for energy resources, such as uranium in Greenland[63] and oil and gas in Spanish (now Western) Sahara. Spain's Saharan oil exploration, however, was heavily dependent on cooperation with mostly American oil companies.[64] The Netherlands arguably had an advantage over other smaller nations, not only in terms of the fabulous oil riches of the Dutch East Indies, but also because Royal Dutch Shell was a largely Dutch company with great muscle in the international arena. It is difficult to judge whether Shell's market power actually contributed to ease Dutch vulnerability to oil imports, but it is interesting to note that the Netherlands Antilles came to host a key refinery in which Shell processed Venezuelan oil during the interwar era, and that supplies from this refinery played an important role in Dutch oil imports.[65]

Even an Eastern European country such as Bulgaria, which in the literature about the Cold War period was often thought of as being locked up behind the Iron Curtain, set out to cope with oil import dependence by participating in foreign upstream activities. In this case, the main producer country at focus was Libya. The friendship between Bulgaria's political leader Todor Zhivkov and Libyan dictator Muammar al-Gaddafi enabled the Bulgarians to obtain a valuable production-sharing agreement in Libya. This was seen to reduce Bulgarian oil import vulnerability.[66]

The Bulgarian-Libyan experience can be interpreted as a typical example of what analysts nowadays often refer to as *energy diplomacy*, that is, the use of diplomatic channels or simply high-level political friendships to counter vulnerability and strengthen energy security in the geopolitical arena. At the outset of energy's internationalization, foreign-policy makers and diplomats were involved on the margin, if at all; cross-border flows of

Figure 5.2 Bulgaria's political leader Todor Zhivkov and Libyan dictator Muammar al-Gaddafi in the Libyan Desert, 1984. Photo: Bulgarian News Agency.

coal and oil were negotiated and controlled by private companies. From around 1910, however, energy companies increasingly sought to involve state actors to support their foreign energy adventures. They usually did so by *securitizing* energy imports – that is, seeking to convince state actors that the supplies in question were of importance for their country's national security. In other cases, it was the state actors themselves that reinterpreted the significance of fuel flows in strategic terms, demanding a say in the negotiations with exporting nations. Today it is regarded as self-evident that an importing nation's diplomatic representatives assist energy-importing companies in negotiating deals with exporting nations. This is especially so when energy imports are negotiated as part of larger foreign trade agreements, in which the import-dependent country's *exports* – that is, of goods and services that are not necessarily related to energy – are also at stake.

Fostering fruitful *general* relations with exporting nations is also of immense importance in the context of energy's internationalization. When a Swedish prime minister visits his or her Norwegian counterpart, the two do not necessarily discuss Swedish oil imports from Norway; yet their handshakes and friendly discussions play an important part in maintaining the cooperative political relations that, in turn, are a prerequisite for trouble-free energy relations between the two nations. Many oil- and gas-importing nations have shown themselves extremely keen to maintain fruitful political relations with countries such as Russia, Iran and Saudi Arabia – although the importing country's energy interests here often runs counter to basic political convictions in terms of, say, human rights and political freedom. They know that friendly political relations may play an important role both in further negotiations about energy supplies and in solving problems in the existing energy trade. Constructive political relations may open up opportunities for privileged access to scarce energy resources – beyond what pure market mechanisms are able to bring.[67] The 1973 oil crisis, at the very latest, reminded the world's oil importers of the necessity to foster fruitful political relations with the Arab countries. Today it is only very reluctantly that energy importers accept demands for ideologically and politically motivated sanctions against countries such as Russia and Iran.

At the same time, energy importers have found that they can manage vulnerability by cooperating with *each other*. The IEA, originally set up to coordinate oil imports in the case of a major supply crisis, is the most famous example, but there are many others. Sometimes, *ad hoc* customer consortia have been created with participants from several countries, the idea being to strengthen the bargaining power of importers in negotiations with powerful exporters. In the 1970s, for example, gas companies from several European countries joined forces in seeking favourable conditions for imports from Algeria and Norway. Needless to say, the actors on the exporter side have been less enthusiastic about such cartelization by their customers. Sometimes the exporters have simply refused to accept such approaches, insisting on separate negotiations with each customer.

Ad-hoc forms of cooperation among importers have been common in numerous crisis situations. The two world wars, in particular, featured far-reaching cooperation in energy, whereby governments also joined forces with the large international oil companies. Such ad-hoc alliances were repeatedly mobilized in the post-war era as well, specifically during the 1951–1953 Korean War and Iranian nationalization crisis, the 1956–1957 Suez Crisis, the 1967 Six-Day War and, in particular, the 1973–1974 oil crisis. It was only thanks to this cooperation that the supply disruptions at these times did not lead to a total collapse in global flows. In a complex logistical feat, the oil companies managed to move oil around in a sharing system in such a way that most needs could be met globally in spite of embargoes against some countries by major exporters.[68]

Another interesting form of cooperation among energy-importing nations takes the form of sharing system activities such as refining and storage. In the 1970s, for instance, the Netherlands joined forces with Britain and Germany in building a joint uranium enrichment facility. The Dutch would hardly have had the possibility to build such a facility by themselves, but thanks to transnational cooperation they were now able to internalize this critical system activity.[69] As for energy storage, a surprisingly large share of the world's oil, gas and electricity storage capacities are internationalized. Luxembourg stored no less than 89% of its oil stocks abroad as of 2017. Estonia, Ireland, Belgium and New Zealand kept around one-third of their oil stocks in foreign locations.[70] This may seem paradoxical, since storage facilities are often thought of precisely as tools for coping with international dependencies in energy. Yet the high costs of keeping oil and other fuels in storage have motivated small countries to cooperate, both with each other and with larger nations, for example under the auspices of IEA. Natural gas storage is highly internationalized today as well. Sweden, for example, has no domestic gas storage facilities, but relies on Denmark for this service,[71] while Finland, as we have seen, is endeavouring to gain access to existing storage capacities in Latvia. In electricity, the very idea of internationalization is intimately connected with the need for back-up capacities, since it is usually cheaper and more convenient to rely on a foreign grid than on a domestic emergency power plant for help in an emergency situation.

Seeking interdependence

In the preceding chapter, it was noted that while energy importers are dependent on energy exporters, the converse is also true. Hence, it is often useful to discuss international energy relations in terms of *interdependence*. However, interdependence does not come about just like that, in an automatic or predetermined way. More often than not it is the result of deliberate strategizing by importers and exporters, who consciously aim for an interdependent relationship precisely because they believe that it will reduce their own vulnerability. In the case of natural gas, Western European import arrangements from Russia for a long time were arranged through complex countertrade deals, in which Europe's gas imports were balanced through Western European exports of steel pipe, compressor stations and other critical equipment and machinery. Over time, the Russians arguably became more dependent on these supplies than Western Europe on Russian natural gas. Japan's dependence on Chinese oil in the 1980s was similarly balanced through extensive Chinese imports of Japanese technology, which the Chinese were unable to source from any other country at that time. It has also been common for the United States, China, Russia and other great powers to negotiate access to fuel in various developing countries in return for weapons. Needless to say, such deals have been subject to criticism from both moral and military points of view. From an interdependence point of view, however, these forms of countertrade offer considerable advantages for vulnerability management as compared to the alternative of accessing foreign energy through the open market. In the course of the 1990s, market-based relations increasingly came to replace the broader trade deals – especially in oil, where after the oil price collapse in 1986 the spot market became the place where most international energy deals were settled. After the renewed surge and continued turmoil in energy prices in the 2000s, however, some analysts have again started to emphasize the advantages of broad, long-term contracts as a useful mechanism to cope with energy dependence.

Actors have also sought interdependence by placing themselves in "hub" positions in international energy flows – and in the related flows of capital and technology. Many smaller nations, which otherwise are typically thought of as particularly vulnerable in international energy relations, have successfully coped with their energy dependence in this way, skilfully exploiting their particular geographical locations. Switzerland early on established itself as a key hub in the European electricity system, taking on a role as interconnector between the German and Italian national grids – in a way that was widely regarded as beneficial for Switzerland's own supply security. From the 1960s on, Switzerland became a key transit country for oil and gas pipelines as well, while also emerging as the world's most important financial hub in oil.[72] While the Swiss remained almost totally dependent on other countries for its oil and gas supply, its hub role was widely perceived as an effective way to reduce the country's vulnerability to international supply crises. When the Soviet Union concluded major gas export deals with West Germany, Italy and France in the late 1960s and early 1970s, similar hub positions were successfully attained by Austria and Czechoslovakia. For both, being located on the export route between the Siberian gas fields and the larger Western European gas distributors was seen to reduce their vulnerability – even though they remained greatly dependent on imported natural gas in a statistical sense. Today, Gazprom's enthusiasm for gas pipelines under the Baltic and Black Seas is a most prominent example of how powerful actors may be greatly troubled by the key hub roles played by smaller countries and how this can result in a perceived need to free themselves from transit dependencies on "problematic" nations, in this case mainly Ukraine. But small nations also see opportunities in the larger nations' efforts to restructure their energy exports and imports. This is the case, for example, in Azerbaijan, Georgia, Greece and several nations in southeastern Europe, which have become (or hope to become) transit countries for oil and gas on its way from Central Asia to Western Europe.[73]

In the age of renewables, the quest for hub positions is likely to be equally intense. In the early 2000s Rotterdam and Antwerp, for instance, started promoting themselves as hubs in the European biofuels system, emerging as powerful "biomainports" where biofuels and biofuel feedstocks are now both imported and exported on a grand scale. And electricity companies in various countries are already in the midst of an intense struggle for hub positions in yet-to-be-built "supergrids" – in Europe, Asia and elsewhere.

Exercises

- Choose an import-dependent nation and analyse some of its state-led efforts to reduce and/or cope with its energy dependence. Consider both internal and external approaches/methods.
- Take a look at the photo of Todor Zhivkov and Muammar al-Gaddafi. Can you identify further friendships at the highest political level that have allowed energy importers to reduce their vulnerability in the geopolitical arena?
- Discuss, in further depth, how state actors may mobilize renewable energy sources in their attempts to cope with energy import dependence.

Notes

1 For this distinction, see, for example, Duffield, *Fuels Paradise*, 33–41.
2 See, for example, Hölsgens, "Resource Vulnerability and Energy Transitions in the Netherlands".
3 Avango et al., "Swedish Explorers, In-Situ Knowledge and Resource-Based Business in the Age of Empire".
4 Rambousek, *Die ÖMV-Aktiengesellschaft*.
5 Murgescu, *Anything but Simple*.
6 Holmberg, *Survival of the Unfit*.
7 Tchalakov and Mitev, "Energy Dependence behind the Iron Curtain"; Arapostathis and Fotopolous, "Transnational Energy Flows".
8 Camprubi, "Whose Self-Sufficiency?"
9 Tchalakov et al., "Bulgarian Power Relations", 140f.
10 Kaijser, "Striking Bonanza"; Hölsgens, "Resource Vulnerability and Energy Transitions in the Netherlands".
11 Högselius et al., *Europe's Infrastructure Transition*, Chapter 7.
12 Högselius, "The Saudi Arabia of the Far East?"

13 Kaijser and Högselius, "Under the Damocles Sword"; Myllyntaus, *Electrifying Finland*; Haller and Gisler, "Lösung für das Knappheitsproblem oder nationales Risiko?"
14 Arapostathis and Fotopolous, "Transnational Energy Flows".
15 Duffield, *Fuels Paradise*, 197.
16 Ibid., Chapter 7.
17 For example, Cantoni, "Second Galicia?"
18 Högselius et al., "Natural Gas in Cold War Europe".
19 One of the most interesting small-country cases is Finland. See Kuisma, "A Child of the Cold War".
20 Hölsgens, "Resource Vulnerability and Energy Transitions in the Netherlands".
21 Yergin, *The Prize*, 694.
22 Sjölander et al., *Motorspriten kommer*, Chapter 6.
23 For an in-depth account of this development, see, for example, Birkenfeld, *Der synthetische Treibstoff 1933–1945*.
24 Rüdiger, *From Import Dependency to Self-Sufficiency in Denmark*.
25 Camprubi, "Whose Self-Sufficiency?"
26 Fjaestad and Jonter, "Between Welfare and Warfare".
27 Fridlund, *Den gemensamma utvecklingen*.
28 Camprubi, "Whose Self-Sufficiency?"
29 Duffield, *Fuels Paradise*, 158–166.
30 For an in-depth historical account of the US case, see Beaubouef, *The Strategic Petroleum Reserve*.
31 Duffield, *Fuels Paradise*, 119.
32 Ibid., 154. In 1970 the German government further "decided to create a reserve of 8 million tons of oil, about 60 million barrels, or the equivalent of 25 days of imports at that time, although construction of the reserve did not actually begin until two years later."
33 "Council Directive of 20 December 1968 imposing an obligation on member states of the EEC to maintain minimum stocks of crude oil and/or petroleum products", 68/414/EEC.
34 Duffield, *Fuels Paradise*, 200.
35 Ibid., 169.
36 Högselius et al., "Natural Gas in Cold War Europe", 50.
37 Söderholm, "Fossil Fuel Flexibility", 81.
38 Rüdiger, *From Import Dependency to Self-Sufficiency in Denmark*.
39 Högselius et al., "Natural Gas in Cold War Europe", 48.
40 Duffield, *Fuels Paradise*, 173.
41 Camprubi, "Whose Self-Sufficiency?"
42 Yergin, *The Prize*, 268f.
43 Kaijser and Högselius, "Under the Damocles Sword".
44 Rüdiger, "From Import Dependency to Self-Sufficiency in Denmark".
45 Kaijser and Högselius, "Under the Damocles Sword".
46 Rüdiger, "From Import Dependency to Self-Sufficiency in Denmark".
47 Hölsgens, "Resource Vulnerability and Energy Transitions in the Netherlands".
48 See Åberg, "A Gap in the Grid".
49 Camprubi, "Whose Self-Sufficiency?"
50 Richter and Holz, "All Quiet on the Eastern Front?"
51 Arapostathis and Fotopoulos, "Transnational Energy Flows".
52 Duffield, *Fuels Paradise*, 140.
53 Ibid., 158.
54 For example, Klare, *Rising Powers, Shrinking Planet*, 73ff.
55 For the "Greater Iraq" case, see Yergin, *The Prize*, 771.

56 See, for example, "US Warships Set to Begin Escorts of Gulf Tankers", *New York Times*, 22 July 1987.
57 See, for example, "China Seeks Oil Security with New Tanker Fleet", *Oil and Gas Journal*, 9 October 2006.
58 Klare, *Rising Powers, Shrinking Planet*.
59 Klare, "Petroleum Anxiety and the Militarization of Energy Security", 47.
60 "Blood for Uranium: France's Mali Intervention Has Little to Do with Terrorism", *Ceasefire*, 17 January 2013.
61 Rahim, "Regional Issues and Strategic Responses", 96.
62 Avango et al., "Swedish Explorers, In-Situ Knowledge and Resource-Based Business in the Age of Empire", 328ff.
63 Nielsen and Knudsen, "Too Hot to Handle".
64 Camprubi, "Whose Self-Sufficiency?"
65 Hölsgens, "Resource Vulnerability and Energy Transitions in the Netherlands".
66 Tchalakov and Mitev, "Energy Dependence behind the Iron Curtain".
67 See, for example, Leverett, "Resource Mercantilism".
68 Yergin, *The Prize*.
69 Schrafstetter and Twigge, "Spinning into Europe".
70 IEA, "Closing Oil Stock Levels in Days of Net Imports".
71 Högselius et al. "Natural Gas in Cold War Europe", 50.
72 Haller, "Globale Geschäfte".
73 Cornell et al., "Geostrategic Implications of the Baku-Tbilisi-Ceyhan Pipeline"; Arapostathis and Fotopolous, "Transnational Energy Flows".

Further reading

Duffield, John S. *Fuels Paradise: Energy Security in Europe, Japan, and the United States*. Baltimore, MD: Johns Hopkins University Press, 2015.

Högselius, Per, Anique Hommels, Arne Kaijser and Erik van der Vleuten, eds. *The Making of Europe's Critical Infrastructure: Common Connections and Shared Vulnerabilities*. Basingstoke and New York: Palgrave Macmillan, 2013.

Moran, Daniel, and James A. Russell, eds. *Energy Security and Global Politics: The Militarization of Resource Management*. London and New York: Routledge, 2009.

Bibliography

Åberg, Anna. "A Gap in the Grid. Attempts to Introduce Natural Gas in Sweden 1967–1991". PhD thesis, KTH Royal Institute of Technology, 2013.

Arapostathis, Stathis, and Yannis Fotopolous. "Transnational Energy Flows, Capacity Building and Greece's Quest for Energy Autarky, 1914–2010". *Energy Policy*, (forthcoming).

Avango, Dag, Per Högselius and David Nilsson. "Swedish Explorers, In-situ Knowledge and Resource-Based Business in the Age of Empire". *Scandinavian Journal of History* 43, 3 (2018): 324–347.

Beaubouef, Bruce A. *The Strategic Petroleum Reserve: US Energy Security and Politics, 1975–2005*. College Station: Texas A&M University Press, 2007.

Birkenfeld, Wolfgang. *Der synthetische Treibstoff 1933–1945: Ein Beitrag zur nationalsozialistischen Wirtschafts- und Rüstungspolitik*. Göttingen: Musterschmidt-Verlag, 1964.

Camprubi, Lino. "Whose Self-Sufficiency? Energy Dependency in Spain from 1939". *Energy Policy*, (forthcoming).

Cantoni, Roberto. "Second Galicia? Poland's Shale Gas Rush through Historical Lenses". *Geological Society, London, Special Publications* 465 (14 May 2018): 201–217.

Cornell, Svante E., Mamuka Tsereteli and Vladimir Socor. "Geostrategic Implications of the Baku-Tbilisi-Ceyhan Pipeline". *Oil, Gas & Energy Law* 4 (2006).

Duffield, John S. *Fuels Paradise: Energy Security in Europe, Japan, and the United States*. Baltimore, MD: Johns Hopkins University Press, 2015.

Fjacstad, Maja, and Thomas Jonter. "Between Welfare and Warfare: The Rise and Fall of the 'Swedish Line' in Nuclear Engineering". In *Science for Welfare and Warfare: Technology and State Initiative in Cold war Sweden*, edited by Per Lundin, Niklas Stenlås and Johan Gribbe. Sagamore Beach: Science History Publications, 2010.

Fridlund, Mats. *Den gemensamma utvecklingen: Staten, storföretaget och samarbetet kring den svenska elkraftstekniken*. Stockholm: Symposion, 1999.

Haller, Lea. "Globale Geschäfte: Wie die ressourcenarme Schweiz zur Drehscheibe für den globalen Rohstoffhandel wurde". *NZZ Geschichte* 4 (2016): 80–95.

Haller, Lea, and Monika Gisler. "Lösung für das Knappheitsproblem oder nationales Risiko? Auf Erdolsuche in der Schweiz". *Berichte zur Wissenschaftsgeschichte* 37 (2014): 41–59.

Högselius, Per. *Red Gas: Russia and the Origins of European Energy Dependence*. Basingstoke and New York: Palgrave Macmillan, 2013.

Högselius, Per. "The Saudi Arabia of the Far East? China's Rise and Fall as an Oil Exporter". *The Extractive Industries and Society* 2 (2015): 411–418.

Högselius, Per, Anna Åberg and Arne Kaijser. "Natural Gas in Cold War Europe: The Making of a Critical Infrastructure". In *The Making of Europe's Critical Infrastructure: Common Connections and Shared Vulnerabilities*, edited by Per Högselius, Anique Hommels, Arne Kaijser and Erik van der Vleuten, 27–61. Basingstoke and New York: Palgrave Macmillan.

Holmberg, Rurik. "Survival of the Unfit: Path-Dependence and the Estonian Oil Shale Industry". PhD thesis, Linköping University, 2008.

Hölsgens, Rick. "Resource Vulnerability and Energy Transitions in the Netherlands since the Mid-Nineteenth Century". *Energy Policy* (forthcoming).

IEA. "Closing Oil Stock Levels in Days of Net Imports". www.iea.org/netimports (accessed 25 January 2018).

Kaijser, Arne. "Striking Bonanza: The Establishment of a Natural Gas Regime in the Netherlands". In *Governing Large Technical Systems*, edited by Olivier Coutard, 38–57. London: Routledge, 1999.

Kaijser, Arne, and Per Högselius. "Under the Damocles Sword: Managing Swedish Energy Dependence in the 20th century". *Energy Policy* (forthcoming).

Klare, Michael T. *Rising Powers, Shrinking Planet: The New Geopolitics of Energy*. New York: Metropolitan Books, 2008.

Klare, Michael T. "Petroleum Anxiety and the Militarization of Energy Security". In *Energy Security and Global Politics: The Militarization of Resource Management*, edited by Daniel Moran and James A. Russell, 39–61. London and New York: Routledge, 2009.

Kuisma, Markku. "A Child of the Cold War: Soviet Crude, American Technology and National Interests in the Making of the Finnish Oil Refining." *Historiallinen Aikakauskirja* 2 (1998): 136–142.

Lagendijk, Vincent. *Electrifying Europe: The Power of Europe in the Construction of Electricity Networks*. Amsterdam: Aksant, 2008.

Leverett, Flynt. "Resource Mercantilism and the Militarization of Resource Management: Rising Asia and the Future of American Primacy in the Persian Gulf". In *Energy Security and Global Politics: The Militarization of Resource Management*, edited by Daniel Moran and James A. Russell, 211–242. London and New York: Routledge, 2009.

Murgescu, Bogdan. "Anything but Simple: The Case of the Romanian Oil Industry". In *History and Culture of Economic Nationalism in East Central Europe*, edited by Helga Schultz and Eduard Kubu, 231–250. Berlin: Berliner Wissenschafts-Verlag, 2006.

Myllyntaus, Timo. *Electrifying Finland: The Transfer of a New Technology into a Late Industrialising Economy*. London: Macmillan, 1991.

Nielsen, Henry, and Henrik Knudsen. "Too Hot to Handle: The Controversial Hunt for Uranium in Greenland in the Early Cold War". *Centaurus* 55 (2013): 319–343.

Rahim, Saad. "Regional Issues and Strategic Responses: The Gulf States". In *Energy Security and Global Politics: The Militarization of Resource Management*, edited by Daniel Moran and James A. Russell, 95–111. London and New York: Routledge, 2009.

Rambousek, Herbert. "Die ÖMV-Aktiengesellschaft: Entstehung und Entwicklung eines nationalen Unternehmens der Mineralölindustrie". PhD thesis, Wirtschaftsuniversität Wien, 1977.

Richter, Philipp M., and Franziska Holz. "All Quiet on the Eastern Front? Disruption Scenarios of Russian Natural Gas Supply to Europe". *Energy Policy* 80 (2015): 177–189.

Rüdiger, Mogens. "From Import Dependency to Self-Sufficiency in Denmark, 1945–2000". *Energy Policy* (forthcoming).

Schrafstetter, Susanna, and Stephen Twigge. "Spinning into Europe: Britain, West Germany and the Netherlands – Uranium Enrichment and the Development of the Gas Centrifuge, 1964–1970". *Contemporary European History* 11, 2 (2002): 253–272.

Sjölander, Annika Egan, Helena Ekerholm, Jenny Eklöf et al. *Motorspriten kommer! En historia om etanol och andra drivmedel*. Möklinta: Gidlunds förlag, 2014.

Söderholm, Patrik. "Fossil Fuel Flexibility in West European Power Generation and the Impact of System Load Factors". *Energy Economics* 23 (2001): 77–97.

Tchalakov, Ivan, and Tihomir Mitev. "Energy Dependence behind the Iron Curtain: The Bulgarian Experience". *Energy Policy* (forthcoming).

Tchalakov, Ivan, Tihomir Mitev and Ivaylo Hristov. "Bulgarian Power Relations: The Making of a Balkan Power Hub". In *The Making of Europe's Critical Infrastructures: Common Connections and Shared Vulnerabilities*, edited by Per Högselius, Anique Hommels, Arne Kaijser and Erik van der Vleuten, 131–156. Basingstoke and New York: Palgrave Macmillan, 2013.

Yergin, Daniel. *The Prize: The Epic Quest for Oil, Money and Power*. London: Simon and Schuster, 1991.

6 Energy as a foreign policy tool

Foreign policy and the purpose of energy systems

What is the purpose of the global energy system? Why do we need it? This is a basic – perhaps too basic, even silly – question to ask. Yet the answer is anything but clear-cut, especially when we view energy from a sociotechnical systems perspective.

From a *technical* point of view, it is common to postulate that the purpose of an energy system is to fulfil "the demand for energy services", "to deliver energy services such as illumination, comfortable indoor temperatures, refrigeration, transportation, etc.", "to convert the potential energy in the raw energy resource (fossil fuels, uranium, wind, PV) to a useful energy form, which can be distributed and used freely for any energy service"[1] and so on. But as we saw in Chapter 2, energy systems are not merely technical entities; they are *sociotechnical* constructs. And from a sociotechnical point of view, the purpose – or goal – of an energy system is highly ambiguous. In particular, the sociotechnical perspective forces us to acknowledge that the purpose of the system can lie *outside* of energy supply as such. The goal can be something that is not technical or physical at all.

Moreover, we need to look at energy supply from the perspective of different actors and stakeholders. As emphasized earlier, different categories of actors have different interests. This means that they see different opportunities in taking action in relation to energy – and their goal is to exploit those opportunities. Scientists, inventors and entrepreneurs typically view energy systems as levers of personal riches and fame. The top managers of energy companies also see personal opportunities, while at the same time viewing energy systems as moneymaking machines for their shareholders. Environmental organizations may view new energy systems as vehicles in combatting pollution and climate change. Terrorists view energy systems as potential tools in holy wars. And then we have the state actors, which, as we have seen, are more diverse in their interests than is usually acknowledged.

In this chapter we look specifically at the interest of foreign policy actors. What is the purpose of global energy from a foreign policy point of view? Foreign ministries naturally view energy systems from the perspective of their entanglement with international relations. In Chapter 5, we saw how foreign policy actors are called upon in attempts to cope with energy dependence. But their own prime concerns are with existential foreign policy goals such as ensuring the country's territorial integrity, its ability to prevent wars and solve international conflicts, its reputation in the international arena, its capacity to build political alliances with other countries and so on. And they view global energy as a powerful tool – a metaphorical lever – in seeking to attain these goals. Accordingly, it is energy's perceived impact on wider foreign policy concerns that constitutes the point of departure for their strategizing in the energy field.

Manipulating flows

How, concretely, can governments use energy as a foreign policy tool? In the following, I sketch five basic avenues of action, or ways to "manipulate" energy for foreign policy purposes.

First, governments may aim to *manipulate cross-border flows* of energy. This usually translates into *sanctions* of various kinds. Negative (punishing) sanctions are deployed to put pressure on a foreign government to change its policies or actions in one field or another – which, importantly, is usually unrelated to energy – and are then removed if the foreign government "gives in". Positive (rewarding) sanctions are deployed to endorse a foreign government's behaviour, with the latent possibility that the rewards might be removed if that behaviour changes. Import and export embargoes are the most radical ways to operationalize negative sanctions, with the 1973 Arab oil embargo as an iconic case. The Arab governments involved in the embargo successfully forced a number of countries to rethink their political stance on the Arab-Israeli conflict. Japan was probably the country most seriously affected. At the height of the conflict, Saudi Arabia's long-term oil minister Zaki Yamani explained the situation to the Japanese: "If you are hostile to us you get no oil. If you are neutral you get oil but not as much as before. If you are friendly you get the same as before". The problem was that since the end of World War II, Japan's foreign policy had centred on an alliance with the United States, which meant that Washington expected Tokyo to side with Israel in the Yom Kippur War. Henry Kissinger, newly appointed US Foreign Secretary, flew to Tokyo in November 1973 to persuade the Japanese government that a change in its foreign policy was not necessary. But the Japanese oil industry was extremely dependent on OPEC oil – 77% of Japan's energy consumption was oil, and most of this came from the Middle East. In this situation, the Japanese finally arrived at the conclusion that their political relations with the Arab countries must be prioritized. So, a few days after Kissinger's visit, Tokyo publicly declared that it was changing its foreign policy *vis-à-vis* the Middle East, from now on endorsing the Arab position in the conflict against Israel. That declaration represented "Japan's first major split on foreign policy with the United States in the post-war era". At about the same time, the European Community, too, painfully aware of its high dependence on Arab oil, decided to officially support the Arab position in the Arab-Israeli conflict. In this case the Arabs did not think this move sufficient, but continued to forcefully argue that the Europeans, in their foreign policies, must keep "putting pressure on the United States on Israel".[2]

In the case of natural gas, the political nature of Russia's recurring supply disruptions to several ex-Soviet states in Eastern Europe and the Caucasus have been much debated, although the relationship between these disruptions and the Kremlin's foreign policy ambitions have not been as explicit as in the Arab oil case. The Kremlin has never publicly declared that it has issued a gas export embargo on any country to attain foreign policy goals. But both foreign policy actors in the importing countries and an army of Western analysts have interpreted Russian gas as a foreign policy tool, most markedly so in the case of Ukraine. The much-publicized supply disruptions of 2006 and 2009, in particular, were widely interpreted as attempts to influence Ukrainian foreign policy. What is fundamentally at stake here has been the struggle between Russia and the European Union (EU) – and between pro-Russian and pro-EU actors in Ukraine – for political power over and in Ukraine. The Kremlin regards Ukraine and other former ex-Soviet republics as Russia's "near abroad" and part of its natural "sphere of influence" (to use a classical geopolitical concept). The EU and pro-Western actors in Ukraine, by contrast, view Ukraine as a potential EU member state. And Ukrainian, Russian and EU foreign-policy makers all realize that natural gas can be mobilized as a potentially decisive instrument in this struggle.[3]

Figure 6.1 Europe's dependence on Russian natural gas is often argued as constituting an "energy weapon". This suggestive photo, showing Gazprom's computerized control room in Moscow, conveys the impression that a mouse click is all that's needed to disrupt the massive gas flows from Russia to foreign nations. Photo credit: Science Photo Library.

Another notable embargo case is that of uranium exports to India. Here, the United Nations played a pivotal role in bringing a vast majority of the world's countries together in the Non-Proliferation Treaty (NPT), which was signed in 1968 and entered into force in 1970. The NPT signatories agreed to allow exports of uranium only to other members of the treaty, and members other than those already in possession of nuclear weapons had to relinquish any activities related to the development of such weapons. India, controversially, chose to prioritize developing nuclear arms and hence did not become an NPT member. This led to a ban on uranium exports to India. Given India's poor indigenous uranium resources, many Western governments hoped that the uranium export ban would force India to reconsider its military ambitions in the nuclear field. This failed. India successfully tested its first atomic bomb in 1974, to the outrage of the Western world, but inspiring numerous other developing countries to pursue nuclear-weapons-based military strategies. Eventually, in 2008, the uranium export embargo was loosened. The main reason was that India was now considered an important ally to the Western world in South and Central Asian geopolitics, especially in the fight against terrorism. Since then, India has been able to import uranium from a variety of suppliers. However, uranium exports to India still remain controversial, especially in Australia.[4]

In the context of renewable energy, it has been hypothesized that electricity exporters could similarly make use of the embargo method to attain foreign policy goals. Today most international electricity connections take the form of synchronized AC connections. The interconnected grids are generally interdependent, and electricity may flow in opposite directions depending on factors such as weather, seasons and hourly demand. There are few large-scale net exporters of electricity. However, there is now a trend towards construction of HVDC links, which are often built for the purpose of electricity exports rather than grid stability – and some of these connections are deliberately designed for one-way flows. The first HVDC links were built back in the 1960s. Among others, Finland started importing electricity from Russia through a "back-to-back" HVDC arrangement. That trade has worked well since its start in the 1980s. But in 2011–2012, Finland was suddenly shaken by a number of unexpected reductions in the power flow. There was no sign that Russia was trying to use the HVDC link for foreign policy purposes; rather, the electricity company on the Russian side of the border appears to have diverted some of the "Finnish" electricity to the domestic Russian market when rates were higher there. The goal was to maximize revenues. Yet, the events made the Finnish grid operator reinterpret the HVDC connection as one that might potentially be used to exert pressure on Finnish foreign policy *vis-à-vis* Russia.[5] Similar fears have been debated in the context of the much-publicized Desertec project, for example, which was launched as a grand vision in 2009 and centres on massive solar energy exports from North Africa and the Middle

East to less sunny regions in continental Europe.[6] The grand visions of the Chinese State Grid to export solar and other electricity from sunny Xinjiang to Central Asia and perhaps even to Europe also immediately gave rise to suspicion among Western actors when first announced. Crimea, the Ukrainian Black Sea peninsula that was occupied and then annexed by Russia in 2014, is one of the few places that have actually been affected by politically motivated electricity supply disruptions. In this case, however, state actors played no role; instead, the connection was cut as unidentified saboteurs – presumably radical Ukrainian nationalists – blew up several pylons in southern Ukraine which supported the transmission lines.[7]

Energy exporters have often attempted to combine embargoes with attempts to "divide and rule", seeking to splinter their enemies politically and preventing them from forming a united front. In 1973 the Arab countries shrewdly divided the world's oil-importing nations into several different categories – from "embargoed" (the United States, in particular) to "most favoured" – correctly anticipating that this would make it more difficult for the importers to come to an agreement about sharing the available oil resources with each other.[8] The Soviet Union, in the initial phase of its Cold War gas export campaign, likewise divided the capitalist countries into two groups: one that would be allowed to import natural gas, and one that would not. The Kremlin set out to develop very fruitful energy relations with Italy, Austria, France and Finland – hoping that gas exports to these countries would help to foster cooperation in other fields as well – while refusing to talk with West Germany. German "imperialism" was allegedly "the USA's chief ally in Europe in aggravating world tension", and the federal government in Bonn was seen to be composed of dangerous "former Nazis and even war criminals", as Soviet leader Leonid Brezhnev phrased it at the 1966 Congress of the Soviet Communist Party. The other countries, by contrast, were praised for their constructive and friendly attitude.[9]

It should be emphasized that importers, too, may make use of the embargo tool. They can do so by simply refusing to import energy as a means of putting pressure on the exporting nation to change its actions and policies. Recent examples here include Western sanctions against purchases of Iranian crude oil and imports of North Korean coal. Similarly, many Western nations have refused to import uranium from countries they perceive to be politically dubious, such as Namibia during the apartheid era.[10] A related case is the politically enforced withdrawal of Western-controlled oil companies from countries such as Sudan on the basis of dissatisfaction with human rights issues, freedom of speech, corruption, environmental scandals and the like. The US House of Representatives, for example, voted in 2001 to "bar oil firms with investments in Sudan from raising capital on US markets".[11] The idea has been to put pressure on African governments to carry out political reforms in a more democratic and liberal direction. In reality, this has rarely had the desired effect, and over the past two decades the vacuum created by the Western withdrawal has, instead, been filled by state-owned oil companies from China and elsewhere.

In between exporters and importers, transit countries may attempt to manipulate flows for political leverage. This is so particularly in the case of grid-based energy systems such as electricity or natural gas, and when there are no viable alternative routes. After the dissolution of the Soviet Union, Belarus and Ukraine, in their capacity as transit countries for gas on its way to Central and Western Europe, found that they had substantial potential leverage over Russia, and they have not hesitated to make use of this for foreign policy purposes – a circumstance that is often forgotten in the discussions about Russian uses of the "gas weapon". This leverage exerted by transit countries is a main factor behind Russia's recent – and very expensive – move to build up a totally new transit infrastructure through the Baltic Sea, through which the former Soviet republics are simply circumvented.

That case can be compared with the perceived political risks that US system-builders and state actors voiced half a century ago in the context of proposed Alaskan oil transit through Canada to US refineries. The fears here concerned the potential leverage that the Canadian federal government, state governments and First Nations might attain. Their potential leverage became one of the main reasons for the preference of an alternative transportation route, centring on a Trans-Alaskan Pipeline, as it came to be called, and tanker transport from the Alaskan port of Valdez to the US Pacific Coast or to Asia.[12] In the future, we are likely to see similar debates over transit geographies in the context of the long-distance transmission

Figure 6.2 Constructing the Nord Stream I pipeline. Poland's foreign minister Radoslaw Sikorski famously dubbed Nord Stream the "Molotov-Ribbentrop Pipeline", and other actors criticized the project on environmental grounds. Here the pipeline is being lowered into the Baltic Sea by the Italian vessel Castoro Dieci in summer 2011. Photo: Nord Stream AG.

of renewable electricity. For example, the Desertec project early on led to a discussion about the potential leverage of countries such as Tunisia – already an important transit country for Algerian natural gas – in the envisioned solar electricity supplies from North Africa to Europe.[13]

Manipulating prices

Apart from direct, physical manipulations of energy flows, governments may set out to manipulate flows indirectly through radical price shocks. In the case of easy-to-transport fuels such as oil, coal and uranium, geographically targeted price increases are difficult to implement in a selective manner, since the buyer can often – but not always (see the discussion in Chapter 5!) – turn to another supplier. Hence, for example, the oil price shocks of 1973/1974 and 1979 affected the entire world rather than those countries that the Arab governments regarded as unfriendly. Grid-based systems, by contrast, offer ample opportunities for exerting political pressure through targeted price manipulations. This is because importers have fewer possibilities – and often none at all – to draw upon supplies from elsewhere. Gazprom, to take one of the most extreme examples, has been able to more or less dictate its gas export prices to its customers in a range of ex-Soviet republics – plus Finland – due to the legacy of the Soviet pipeline network, which definitely was not built to enable alternative supplies from elsewhere. As we saw in the previous chapter, gas prices in Eastern Europe vary in a way that in the eyes of many lacks any "natural" logic. Importers have been subject to both price shocks and sudden generous discounts, many of which have been widely interpreted as reflecting the Kremlin's foreign policy goals.

This has been most clearly demonstrated in the case of Ukraine. In December 2013, pro-Russian Ukrainian President Viktor Yanukovich secured a deal with Russia that did not merely guarantee that Russia would continue to ship gas to Ukraine – which was highly dependent on this energy source – but also that this gas would be supplied much more cheaply than before (at a 33% discount). To achieve this, Yanukovich was forced to promise that his government would not sign any political association agreement with the EU. This link between gas prices and foreign policy was perceived as so radical, however, that it triggered massive public protests in large parts of Ukraine. Two months later, Yanukovich was ousted. Gazprom then cancelled the discount, together with other, previous rebates, so that the gas price suddenly increased by 80%! Shortly afterwards, in March 2014, Russian military forces entered and occupied Crimea, and civil war erupted in eastern Ukraine between Russian-supported separatists and the central powers in Kiev.[14]

An important alternative strategy to embargoes and radical price increases is its opposite: to flood foreign markets with cheap energy. For example, when the Soviet Union re-entered the global oil market as an exporter in the mid-1950s, the US government accused the Kremlin of dumping crude in Europe for foreign policy purposes. Washington was upset by large-scale Soviet oil supplies at discounted prices to countries such as Italy, Finland and Sweden, interpreting these deals as attempts by Moscow to wreak havoc in international oil markets, throw a wrench in Western European attempts at political unification and stimulate political support for the Soviet Union among the importing nations.[15] More recently, China has been accused of dumping energy-related items on world markets, especially solar PV cells, but also excess coal. Western analysts have also argued that China sold massive volumes of rare earth elements (REEs) at artificially low prices in the 1990s and early 2000s – in a way that helped to drive Western rare earth companies out of the market; this then paved the way for the much-debated Chinese dominance – and alleged abuses – of rare earth supplies to Japan and elsewhere (see further Chapter 7).

In the world of oil, Saudi Arabia is the only country that has production (and storage) capacity sufficiently large and flexible to be able to practice dumping on a large scale. For this reason, increases and decreases in Saudi oil production are monitored closely, not only by oil companies worldwide but also by governments. Any significant increases in production typically generate suspicions of oil being used by Riyadh as a foreign policy tool. In the 2010s, Saudi production increases have been discussed mainly in the context of competition with North American tight oil. There is no doubt about the fact that Saudi Aramco and the Saudi government – along with other OPEC (and Russian) actors – have at times increased production in an attempt to tamp down world oil prices, the goal being to squeeze the more expensive American oil out of the marketplace. But it is not clear whether this should be interpreted as a business strategy linked to profitability and market share considerations, or rather as a foreign policy tool. We will re-turn later on to the ambiguities involved in assessing situations like this one.

Manipulating system-building

Governments may also *manipulate system-building activities* for foreign policy purposes. In fact, it is rare to see foreign policy actors *initiate* the construction of a new energy system or a new component or link in such a system. However, it is common that they *intervene* in such activities, encouraging or discouraging various projects or trying to reshape them in a way that will serve their interests. The most typical situation here is perhaps one in which a prime minister or a foreign office gets to hear about a project being proposed by an energy company or an equipment supplier. The government may see merit in the project from a foreign policy point of view and decide to support it, sometimes with great enthusiasm, judging that it might help to improve diplomatic relations with the involved countries and strengthen the nation's overall reputation in the international arena. Or it may veto the proposed project on the basis of the perceived political risks. Such considerations have clearly shaped the global energy geography in decisive ways.

During the Cold War, for example, the North Atlantic Treaty Organization (NATO) member states in Western Europe, under pressure from the United States, discouraged the construction of electricity transmission lines across the Iron Curtain – although many leading electrical engineers and international organizations such as the UNECE advocated such links on the basis of both economic and political opportunities.[16] In East Asia, Chinese foreign-policy makers enthusiastically advocated oil export arrangements to Japan in the 1970s, because they hoped that Sino-Japanese energy cooperation would make it less attractive for Japan to develop fruitful political relations with the Soviet Union, an enemy of China at the time.[17] In other cases, energy projects and systems have become bargaining chips in complex political negotiations between two or more states. An interesting case here is Bolivia's (ultimately failed) attempt to build a gas-export pipeline to Chile and conclude a large gas-trade deal with the Chileans; in return, the Bolivian government wished to acquire part of Chile's territory so as to gain direct access to the Pacific Ocean.[18] Another is US state support in 2001 to the controversial oil pipeline – championed by Chevron – that would bring Kazakh oil to world markets by way of Russian territory; while this increased the leverage of Russia in Caspian oil, President George W. Bush decided to support the project in return for Russian assistance in the war it had just launched in Afghanistan.[19]

A popular strategy for manipulating system-building has been to embargo exports of technology and equipment to "unfriendly" states. Such embargoes were first-page news during the Cold War era, when hawkish foreign-policy makers in the United States and elsewhere – mainly NATO countries – sought to prevent technology exports to the Soviet Union, Eastern Europe, China, Cuba and other communist countries. Many energy technologies were "securitized" during that era; they were rhetorically (re)defined as being of national strategic importance and of potential danger if exported to the communist world. Large-diameter steel pipe, for deployment in long-distance oil and gas pipelines, and compressor stations for natural gas systems were among the most hotly debated items, along with turnkey deliveries of entire refineries and gas-processing plants. This development reached two peaks: one in the late 1950s and early 1960s, and the other in the early 1980s. These were difficult times for the Western manufacturers of the items in question, from Britain's Rolls Royce (a leading maker of gas compressors) to Germany's Thyssen and Mannesmann (manufacturers of high-quality steel-pipe). The companies disliked the export restrictions because a country such as the Soviet Union offered one of the world's largest markets for their products.[20] In the twenty-first century we have once again seen a wave of export embargoes in the energy-technology field, with countries such as Russia and Iran as major targets. And once again, the Western manufacturers and service providers are anything but happy about the sanctions.

Energy-related development aid has also played an important role for foreign-policy makers in their attempts to manipulate system-building. Soviet foreign-policy makers, for example, enthusiastically supported the provision of technical assistance and expertise to a range of developing countries. The experts themselves usually regarded their work abroad as serving a higher purpose of world development; they contributed, as they saw it, to eradicating poverty, stimulating industrial growth and supporting societal moderniza-tion. From the Kremlin's point of view, however, the aid programmes served a different purpose. Helping developing countries to harness their energy potential – from hydropower in Egypt to natural gas in Afghanistan – was seen as a way to keep rivalling foreign powers out of these countries and pre-vent the United States from establishing itself as a world political hegemon.

The developed capitalist countries of the world have behaved in more or less the same way. Energy played a central role in post-war Marshall Aid to Europe, which the United States happily provided in order to prevent the spread of communism in Western Europe. Many of the Western European nations that received Marshall Aid later became major donors in their own right, serving a range of developing countries in Africa, Asia and Latin America. Again, energy and other infrastructural sectors were clearly em-phasized. The aid was highly selective in a political sense: it was not given to those countries that were most in need of it, but to countries that were in need of support *and* fulfilled certain political criteria. Sweden, for example, emerg-ing as a major donor in the 1960s, used the lure of access to its world-leading competencies in hydropower construction to endorse and encourage social-ist political reforms in places such as Mozambique and Tanzania.[21] Over time, Western development aid was sometimes withdrawn from countries that did not live up to political expectations. Today Western development aid is being rivalled by Asian initiatives. China, in particular, has become extremely active in numerous resource-rich African countries, often stepping in where Western actors have left in protest against human rights issues and the like. As we have seen, Chinese state-led investments in African oil can partly be interpreted as a vulnerability management strategy in the face of growing oil import dependence. Yet there is no doubt about the fact that these investments – which are now reshaping African energy systems in deci-sive ways – also serve China's foreign policy interests in Africa.

The extent to which foreign policy actors are ready to support or oppose an international energy project is often related to the project's technical, ge-ographical and organizational features. A foreign office may disapprove of a proposed project because it considers its scale too great, or it may support a project under the condition that a different constellation of system-builders take the lead or a different transportation route be selected. In negotiating the post-war petroleum order, for example, Saudi Arabian King Ibn Saud, who was suspicious about everything British, insisted that the planned scaling up of Saudi oil production had to be carried out by American companies only and without British participation.[22] And in the choice of an optimal pipeline route for moving Soviet natural gas to Europe during the Cold War, West German foreign policy *vis-à-vis* East Germany, whose sovereignty as a state Bonn did not recognize until 1972, dictated that no pipelines could cross East German territory – even though that would have been a logical route from an economic point of view.[23] In the twenty-first century, the Indian government has considered it unfeasible, from a foreign policy view, to import piped natural gas from two exceptionally gas-rich countries in its geographical proximity – Iran and Turkmenistan. One reason is that the gas would have to be transited through Pakistan. As a result, the Indian natural gas supply system has, instead, become based on liquefied natural gas (LNG) terminals through which gas can be imported from several different sources (and without the involvement of problematic transit countries) – albeit under much less favourable economic terms than potential pipe-based supplies from Iran or Turkmenistan. South Korea is in a similar situation: it imports massive volumes of LNG from Southeast Asia and the Middle East, although it would probably be economically more efficient to draw on pipeline-based Russian supplies. At the time of writing, the latter option is still considered unacceptable because the gas would have to be transported through North Korea. A final example concerns the "unnatural" choice of pipeline route for moving Caspian oil to markets; it goes through Azerbaijan, Georgia and Turkey, whereas a much more logical choice, from a technical-economic point of view, would have been a Trans-Iranian route. American foreign policy officials, in particular, considered the latter route geopolitically unacceptable. Moreover, the selected route fits in with broader strategic interests of the United States in the Caucasus and Turkey. As Secretary of Energy Bill Richardson put it in November 1998, the choice of pipeline route was to a large extent about "preventing strategic inroads by those who don't share our values".[24]

A special category of investments in which foreign policy motives typically play important roles are projects in geopolitically sensitive or disputed areas, such as border regions or territories where the borders have not yet been settled. Hydropower projects in international river basins belong to this category. In the interwar years, for example, France embarked on a radical project to exploit the hydropower potential of the Upper Rhine. The engineering designs relied on water being transferred from the Rhine into a lateral canal – the Grand Canal, as the French called it – essentially moving the entire river into France. In neighbouring Germany, the project was widely interpreted as a hostile initiative designed to destroy Baden-Württemberg's agriculture. Some German scientists believed that the project would pave the way for the "steppification" of southern Germany, and as such it became an argument for Hitler to go to war with France.[25]

Since then numerous other hydropower projects in international river basins have become subject to heated political disputes. Some concern rivers that, like the Rhine, form the border between two or more countries. The Gabčíkovo–Nagymaros Dam on the Danube between Hungary and what is now Slovakia is one example. It was initiated in 1977 but has not been finished. Hungary suspended ongoing construction of the dam in its part of the river in the context of communism's collapse in the late 1980s, citing harmful environmental consequences. Slovakia wished to proceed, and eventually did so by diverting the river in a controversial way – not unlike the French had done with the Rhine decades earlier. The project has subsequently become a stumbling block in the wider context of Slovakia's and Hungary's troubled political relations.[26]

Other hydropower controversies concern the relations between upstream and downstream nations. When an upstream nation decides to build a hydropower dam in a river, this inevitably has effects on downstream water flows. International legislation and conventions have been developed to smooth over the conflicts such situations present, but this has not prevented some projects from causing diplomatic quarrels, some of them with existential overtones. Some dams have been interpreted as potential "energy weapons" that grant the upstream government substantial political leverage over downstream governments. Critical agricultural interests are often at stake, the nightmare being that the upstream nation either close the dam, thereby temporarily disrupting the water flow, potentially causing a catastrophic drought, or open it in a literal flooding attempt. Examples include Turkey's hydropower visions along the Tigris and Euphrates (with Syria and Iraq as downstream states), and Ethiopia's grand hydropower projects along the Blue Nile (which flows to Egypt).

Furthermore, many governments have sought to mobilize energy-related activities as a tool to support territorial claims and to gain political influence abroad. This builds on a long historical tradition established in the era of imperial expansion, where evidence of economic or industrial activity was widely regarded as a prerequisite for international recognition of territorial claims. In the early twentieth century, for example, Swedish, Norwegian and Russian foreign-policy makers actively supported the establishment of coal-mining enterprises in the Arctic archipelago of Spitsbergen. They did so because they wished to influence the future legal status of Spitsbergen, and hoped that the mines would help them strengthen their position in international negotiations on this matter.[27] In our own times, the British government's decision to allow – and support – oil and gas prospecting and exploration in the waters around the Falkland Islands may also be interpreted as a way for Britain to strengthen its claim for sovereignty in this area, which is contested by Argentina.

In a similar vein, recent conflicts over sea borders and exclusive rights to sub-sea energy resources in areas such as the Arctic Ocean and the East and South China Seas are strongly related to foreign policy – although in this case it is sometimes difficult to discern the extent to which offshore energy resources have been mobilized for foreign policy purposes, or whether the

Figure 6.3 Swedish coal production in Spitsbergen, 1918. The picture shows the first load of coal produced from the Svea mine. Swedish foreign-policy makers supported this business-led mining project, making use of it as a tool in the struggle against other nations for political control of this Arctic archipelago. Photo credit: Swedish National Museum of Science and Technology.

energy supplies as such are at stake. Russia's state-supported activities in the Arctic Ocean are a case in point. On the one hand, these activities may be interpreted as a Kremlin-led effort to assert Russia's political and military dominance in the region. On the other, they must be seen in the context of rapid depletion of Russian onshore oil and gas deposits – and the resulting need to secure additional supplies to sustain fuel exports. State-controlled industry actors such as Gazprom and Rosneft have often been reluctant to move ahead with concrete investments in Arctic exploration, pointing to the enormous technical and economic challenges. But such reluctance may reflect attempts by the companies to secure state subsidies for prospecting and exploration and may in this sense conceal their actual enthusiasm. Yet the impression is that Arctic exploration is being pushed forward primarily by the Kremlin, which, apart from eyeing the region's strategic significance, uses the heroic quest for Arctic energy to compensate for the country's lost superpower status. The planting of a Russian flag on the ocean floor at the North Pole in August 2007 can be seen as an expression of this.

Manipulating ownership and control

Ownership and control are clearly central themes when it comes to energy in the foreign policy context. Foreign ownership over local energy systems has often been highly controversial precisely because it has been perceived as linked to foreign governments' agendas – even if the foreign investors are privately owned. A foreign investor in energy must always prepare itself for situations where nationalist and anti-foreign sentiments suddenly run high, with sabotage, harassment and expropriation of the investments as a real possibility. This became obvious to the international oil majors during the heydays of decolonization in the mid-twentieth century, when a wave of nationalization swept across the oil-producing developing countries, from Mexico to Libya, and more recently to the Spanish oil and gas company Repsol, whose control of Argentina's hydrocarbon flagship YPF was thwarted by President Cristina Fernández de Kirchner's decision to (re) nationalize that company in 2012.

Suspicions against alleged political motives behind foreign acquisitions have sometimes prevented foreign investments from materializing. In the United States, the 2005 Unocal affair is a case in point. Unocal was an American oil and gas company founded back in 1890. In the early 2000s, it gradually became clear that the company was for sale. But who would buy it? Chevron, one of the supermajors and the second largest US-based oil company, emerged as the main contender. It came up with what seemed to be a highly competitive bid, $16.5 billion, far above all others. But then a new actor suddenly entered the game: the Chinese state-owned company CNOOC. It offered to pay even more for Unocal, $18.5 billion. It was the largest Chinese bid ever seen for a US company. Many Americans reacted vehemently against the Chinese offer. The opponents cited national security interests: since CNOOC was a state-controlled company, analysts suspected that the bid might somehow be part of a Chinese government strategy. It was feared that Beijing might use Unocal to achieve certain foreign policy goals. Unocal in Chinese hands, or so the argument went, would be a Trojan horse, of great potential danger to the United States. Underlying the whole affair was, of course, the fact that China was a communist power and, above all, a rapidly rising power on the world political arena. In the end, the takeover attempt failed. A coalition formed by the company's managers, Washington lobbyists and republican politicians managed to introduce a piece of new legislation that prevented CNOOC from buying Unocal.[28]

The Chinese have been more successful in Africa. As we have seen, investments by Chinese state-controlled companies in countries such as Angola and Sudan can largely be interpreted as a way for Chinese oil companies to cope with their growing oil-import dependence. However, the Chinese government and the Communist Party support these investments not only for the sake of Chinese energy security; they also view the companies' African investments as excellent opportunities to build fruitful diplomatic relations with a range of African governments, spread the Chinese model of social and economic development to the continent, and thus also shape Africa's political future.

In Europe, Russia's Gazprom has made numerous important foreign acquisitions. The company has been very eager to strengthen its presence in its customer countries and the countries through which Russian gas is transported. For example, Gazprom now owns the Belarusian and the Armenian gas transport infrastructure and the German gas distributor Wingas. It also holds between 25% and 50% of the main gas companies in Serbia, Moldova, Hungary, Latvia, Estonia and Finland.[29] Gazprom motivates its foreign investments on purely business-strategic grounds. It is simply seeking downstream integration, or so the argument goes, in the same way that energy producers across the world have done for over a century. The company's managers argue that its activities have nothing to do with Russian foreign policy. In the target countries, however, the acquisitions have often evoked perplexity and dismay. The politicians in charge of approving the ownership changes have typically been accused of accepting bribes or of being biased in their support through personal friendships with key Russian managers or decision-makers. Many analysts have interpreted Gazprom's inroads as attempts by the Kremlin to use the company as a foreign policy tool.

Smaller countries have also been active in foreign acquisitions of energy companies. The electricity and district heating systems of Stockholm, the Swedish capital, are currently controlled by the Finnish state-controlled company Fortum. As a result, heat and electricity supplies to the Swedish Government Offices are at the mercy of the Finns! Fortum also has major stakes in several Swedish nuclear power plants. The Swedish state, for its part, through its Vattenfall energy company, controls the electricity systems of key German cities such as Hamburg and Berlin along with most of the former East Germany. The Swedish acquisitions in Germany have sometimes been compared to the aggression of the Swedish army under Gustavus Adolphus during the Thirty Years War (1618–1648), which caused much havoc in the German lands. At that time, the Swedish king was sometimes celebrated as a liberator of the Germans from Catholicism. Today, the Swedish Green Party wants the government to use Vattenfall's power in Germany to liberate Germany from fossil fuels.[30]

Manipulating discourses

Of decisive political importance in the information age is not only what governments *do* but also what they *say* and how they say it. The geopolitics of energy is no exception here. Discourses about energy can become powerful tools in the hands of foreign-policy makers.

Many international energy relations are only marginally linked to foreign policy issues, and for this reason they rarely become subject to public debate. Most energy-related conflicts and crises have their roots in nonpolitical disagreements such as differing interpretations of politically irrelevant contractual clauses, technical problems in the international energy trade, competitive rivalry of a non-political nature and the like. Although such issues may fuel major debates among engineers, lawyers and economists, there is rarely any intrinsic need for political intervention.

However, foreign policy actors sometimes enter such debates by deliberately *politicizing* them. In the same way as terrorist groups "take responsibility" for attacks in which they actually played no role, foreign policy actors may take credit for international energy crises, conflicts and breakdowns that were actually brought about without much political involvement. For example, in the early 1980s, several Western European countries were in the process of negotiating scaled-up imports of Soviet natural gas. Economic recession, however, gradually lessened the European interest in further imports from the East, and by late 1982 gas companies in countries such as Italy and France had opted to take a "pause for reflection" in the negotiations with Moscow. Then, in mid-December 1982, the democratic Solidarity movement in communist Poland was brutally crushed by the military, fuelling political protest all over Western Europe. Most Western analysts concluded that the Soviet Union was most certainly behind this course of events. The Western governments then solemnly declared that they were postponing the gas negotiations with the Soviet Union for political reasons – as a way to show their dislike with the Soviet behaviour. In other words, the "pause for reflection" was politicized.[31]

Another revealing case concerns Algeria's LNG exports to the United States in the 1970s. When the Arab oil embargo was launched in autumn 1973, the Algerian state company Sonatrach was facing a number of technical problems at its LNG export terminals. The Algerians were working hard to solve the problems, but their resolution took time. Then, in December 1973, Algeria's Minister of Energy Belaid Abdessalam reframed the delivery problems as being politically motivated, telling American journalists that "future shipments of LNG to the US on a continuous basis may depend on the satisfactory settlement of the Arab-Israeli conflict".[32] This example points to a confusing conflation of technical and political factors – and to the possibilities that arise for discursively reframing energy-related events in the foreign policy context.

A third case concerns the conflict between Russia and Estonia over natural gas supplies after the Soviet Union's collapse. As of 1993, many ex-Soviet republics found themselves unable to pay for gas deliveries from Russia, and Gazprom was in deep trouble, as it was unable to enforce payments for deliveries. Huge debts were mounting. This development coincided with a diplomatic crisis between Russia and Estonia. The Estonian government had issued a new citizenship law which was seen to hurt ethnic Russians in Estonia. The Kremlin protested vigorously against the law, and in this context it also supported a gas supply disruption to Estonia. However, similar supply disruptions were experienced at that time by other ex-Soviet republics as well. Supplies to Lithuania, for example, were cut off. The Lithuanian government had not issued any anti-Russian laws. However, the country's gas debt was much larger than Estonia's. The impression is that the Estonian disruption would have occurred even in the absence of the citizenship law, but that the Kremlin made use of the gas debt debate to strengthen its activities in the foreign policy arena.[33]

It should be emphasized that discourses may also be manipulated in positive ways. Governments are often eager to "bless" international energy projects that were not actually developed to serve political interests. Top politicians from multiple countries usually take part in inauguration ceremonies in connection with start-ups of transnational hydropower plants, oil and natural gas pipelines, offshore wind power parks in border-near regions and the like – even if the governments actually played only negligible roles in shaping the projects in question. They rarely miss the opportunity to frame international energy projects rhetorically as symbols of international friendship.

Moreover, renewable energy investments in many countries are currently used rhetorically by governments as evidence of responsible environmental thinking and technological innovativeness. The aim is typically to build up a reputation that in turn will attract foreign capital – and, more generally, to strengthen the country's prestige in the international arena. Germany's famous *Energiewende* is a case in point. The energy transition, featuring massive investments into wind and solar energy, has served well to raise Germany's overall international reputation far beyond the energy realm. German politicians are in dire need of such positive discourses, given the fact that in many parts of the world their country is still associated with a dark and violent past.

Energy weapons real and imagined

Whether a certain country possesses an "energy weapon" and if so, whether its government wields this weapon in actual practice, is often subject to debate. Different stakeholders and analysts typically arrive at different interpretations and draw different conclusions regarding the causal relationship between foreign policy and critical events such as a sudden disruption in the cross-border flow of energy, a price shock or the refusal of a foreign partner to cooperate in a new international energy project. The true motives of foreign governments often remain unknown, and even when a government actually attempts to make use of energy for foreign policy purposes, it does not necessarily publicize this. Conversely, a government may announce publicly that it aims to make use of energy as a foreign policy tool, whereas the real motives are of an unpolitical nature. Moreover, it is common that political, economic and other motives are combined or intersect, making it difficult to discern the foreign policy dimension in an energy project.

Analytically, it is often useful to think of an energy weapon as existing only insofar as it is believed to exist, and to place perception rather than objective reality at the centre. Whether or not Russian natural gas has "actually" constituted an energy weapon, for example, the actors involved have been forced to take a position related to the weapon's *perceived* existence. Importantly, Russia's energy weapon, whether "real" or "imagined", has thereby had a very concrete influence on system-building activities such as the dimensioning of Western Europe's underground gas storage facilities, its efforts to build interconnecting pipelines with alternative gas suppliers and its overall ambitions to diversify supply. When the East-West gas trade started in the years around 1970, Western Europeans were highly suspicious of Moscow's intentions, and all importers regarded politically motivated supply disruptions and aggressive price dumping as a real risk when negotiating with the Soviets and constructing the import infrastructure. Huge investments were made in technical facilities whose purpose was to reduce the adverse impact of unexpected Soviet moves. Whether or not the Soviet gas weapon "actually" existed, its socially constructed reality thus had a very tangible impact on the physical characteristics of the European gas system.[34]

The same goes for China's perceived leverage in REEs, which, as we have seen, are critical to scaling up renewable energy systems. In autumn 2010, a Chinese trawler famously collided with several Japanese coast guard vessels near the disputed Senkaku Islands. The Japanese, shocked by the event, captured the Chinese trawler's captain. The event triggered a diplomatic crisis between the governments of the two countries. Shortly afterwards, China announced that it was cutting its REE exports by 40%. World market prices skyrocketed, and Japan was the country most severely hit, since the lion's share of China's REE exports went there. Most Japanese and Western analysts interpreted this as a clear case of China using its "rare earths weapon" against Japan in the context of the imprisoned Chinese skipper and the territorial dispute. However, Beijing firmly rejected the accusation that it was trying to use REE exports to put political pressure on Japan. Instead, it referred to the rapidly growing domestic demand for REEs and the need to scale back production to protect the environment. The important point, however, is that, whether or not China actually wielded its REE weapon, the rest of the world perceived that it had, and foreign actors responded to these perceptions rather than to any objective reality. Hanna Vikström, an expert on the subject, notes that Japan, being alarmed by the affair,

> began to stockpile rare earths and attempted to substitute other metals for rare earth elements, fearing what the future might bring. The government initiated discussions with Vietnam and Mongolia about new production facilities and supplies of these critical metals. Major industrial companies, such as Toyota and Hitachi, for their parts, attempted to eliminate the rare earths contained in their products. The government even began to study the potential for deep water mining of rare earths, seeing it as an unexplored frontier which they could control.

The US Department of Energy, in a similar vein, set out to create a "strategy to increase United States production, find substitute materials and use rare earths more efficiently", and the Pentagon initiated a "complete study of the United States' military dependency".[35]

Appropriation

Historically, foreign policy concerns have rarely constituted the *main* driving force behind energy's internationalization. It is much more common that the main focus is on supply, economic and environmental issues. As a rule, the main visionaries and initiators of transnational energy projects are found within energy companies or at ministries with responsibility for the economy, industry, foreign trade, energy or environment. Foreign ministries, by contrast, are rarely in a position to come up with feasible visions for the creation of new cross-border oil or gas pipelines or electricity transmission lines, new tanker routes, hydropower plants situated on border-defining rivers or the like. Such projects depend on specialized knowledge and competencies that are only available elsewhere in the state administration and in the private sector. In cases where foreign ministries do come up with new visions, they are often ridiculed as unrealistic and naïve by energy experts.

However, once a vision has been presented by other actors, foreign ministries may take an interest in them, assessing them in terms of their likely impact in the foreign policy realm. If they evaluate the project positively, they may become enthusiastic supporters of the project in question and may in that context also contribute – sometimes decisively – to speeding up negotiations at company or government levels. A striking example is the Clinton administration's enthusiasm for oil and gas in the Caspian region. In the immediate post-Soviet era, American interests here were spearheaded by the large American oil companies, with Chevron taking the lead. Chevron was already seeking business opportunities in Kazakhstan in 1990, when the republic was still part of the Soviet Union. Following Kazakh independence in 1991, Chevron forged a deal with Nursultan Nazarbayev's government to exploit Kazakhstan's oil resources, and subsequently initiated the formation of the Caspian Pipeline Consortium in order to move Kazakh oil to markets (through Russian territory). Exxon Mobil, ConocoPhillips and other American firms soon followed suit. Like Chevron, these actors were driven by their entrepreneurial spirit and the lure of profits. Over time, however, the rapidly growing business activities of the companies attracted the attention of Washington, which came to the conclusion that the presence of American companies in the Caspian oil industry could be used for foreign policy purposes. The Clinton administration judged that Chevron and the other companies could help to turn the region into "a bulwark against any future imperial Russian superpower". By helping Kazakhstan and other Caspian nations to launch new energy projects, Clinton's officials believed, "US firms could help generate fresh income for the struggling young Caspian states – thus enabling them to escape the political and economic embrace of Moscow". Against this background it is not surprising that Washington became extremely active in helping American oil companies in their attempts to build fruitful links with local politicians and state actors in the region.[36]

West Germany's Cold War imports of Soviet natural gas are another case in point. This project was championed by regional state and private actors in Bavaria and by the German steel industry, which hoped to sell large volumes of high-quality steel pipe to the Soviet Union in return for the imported fuel. The Federal Ministry of Economy was also involved at an early stage, as was Germany's dominant gas company, Ruhrgas. Once negotiations on a concrete deal began, however, the project captured the attention and imagination of Foreign Minister Willy Brandt, who knew nothing about natural gas, but was about to forge a new German foreign policy *vis-à-vis* the communist countries, based on the concept of "change through rapprochement" (*Wandel durch Annäherung*). Brandt and his close advisor Egon Bahr concluded that the gas negotiations with the Soviet Union could become a powerful instrument in implementing the new policy. The East-West natural gas trade, although it was worked out by other private and state actors, was thus appropriated by the foreign-policy makers for their specific purposes.[37]

Beyond state actors

Before we conclude this chapter, a few words should be said about actors other than national governments that seek to use international energy systems for political purposes. We have already seen how, for example, regional Bavarian actors mobilized transnational oil and gas pipelines in its political and economic struggle with northern Germany in the 1960s and 1970s, and how, more recently, the Bedouin of the Sinai Desert sabotaged Egyptian exports of natural gas to Israel in protest against ethnic discrimination on the part of the central Egyptian government. Separatist groups have also repeatedly made use of energy in their struggles, especially in Africa. In Ethiopia, for example, the Ogaden National Liberation Front in 2007 "overran a Chinese-operated oil field, killing nine Chinese oil workers and abducting six".[38] And in Nigeria, the Movement for the Emancipation of the Niger Delta (MEND) has long wrought havoc on the large Western oil companies that are active in the area by destroying oil pipeline and pumping stations and kidnapping oil-industry employees; the underlying motivation here has been dissatisfaction with the massive environmental destruction caused by Shell and other Western companies, and the failure of the Nigerian government to make use of the country's oil wealth to eradicate poverty.[39]

Transnational terrorist organizations form another conspicuous actor category of relevance here. Although they do not represent any state interests, they nevertheless typically pursue very outspoken foreign policies. Following the 9/11 terrorist attacks in the United States and the US-led military assaults on Afghanistan and Iraq, some analysts feared that Al-Qaeda might plan a major terrorist attack on key oil-system facilities in the Persian Gulf. The main goal of such an attack, they reasoned, would be to attain political ends such as the retreat of the United States from the Middle Eastern region.[40] The pessimists have feared that jihadist uses of energy infrastructures is inevitable in view of the fact that the world's major Muslim populations – the main recruitment base for Islamist terrorist organizations – largely overlap geographically with the locations of the world's largest hydrocarbon reserves. However, some analysts argue that Al-Qaeda and other Islamist terror groups are not very interested in targeting energy systems, noting that "the petroleum-producing infrastructure of the Middle East offers little in the way of worthy symbolic targets". On the fifth anniversary of the 9/11 attacks, Osama bin Laden's deputy, the Egyptian cleric Ayman al-Zawahiri, encouraged jihadists to commence attacks against the energy infrastructures of companies whose revenues went to the "the enemies of Islam". But "despite the unsuccessful suicide car-bombings of some oil facilities in Yemen two days later, there has been a marked absence of such attacks throughout the Arab Muslim world".[41] On the other hand, as suggested by Duane Chapman, even if the actual impact of terrorism on the Middle East's oil system in the early 2000s was negligible in the sense that it has not affected world energy markets, the few attacks that did occur – such as a May 2004 attack on a Saudi Arabian compound housing oil-company personnel – "have done much to compound the political uncertainty" created by the war in Iraq.[42] In this sense, oil terrorism can be said to actually have served the political ends of Al-Qaeda.

Needless to say, over the years, many terrorist and guerrilla groups have also made use of energy to *finance* their violence. Yet, the importance of energy and, in particular, oil as sources of revenue for such groups has probably been exaggerated in media reports. Keeping oil-production systems in operation is an immense technical and managerial challenge, and the rebels are not necessarily capable of dealing with this complexity. A recent study of oil extraction in the territories that ISIS came to occupy, for example, shows that production essentially ceased following the occupation.[43]

Exercises

- Pick a country and try to discern how its government makes – or tries to make – use of energy for foreign policy purposes.
- Collect evidence and arguments that support the claim that the Russian government is using energy as a foreign policy tool. Then try to find counterarguments that point to Russian energy exports as being driven by business rather than politics. Critically estimate the trustworthiness

of your sources. Discuss whether it is possible to come to a definite conclusion regarding the actual extent to which Russian energy exports are
primarily shaped by political as opposed to economic factors.
• Identify at least one additional case where state actors appear to have
appropriated economic or technical conflicts over international energy
flows for foreign policy purposes.

Notes

1 Schrattenholzer et al., *Achieving a Sustainable Global Energy System*, 169;
   IAEA, *IAEA Tools and Methodologies for Energy System Planning*, 7; Storm
   van der Leeuwen, "Nuclear Power".
2 Yergin, *The Prize*, 628f.; Duffield, *Fuels Paradise*, 205.
3 Högselius, *Red Gas*.
4 For example, "Australia's Proposed India Uranium Deal Given Cautious Green
   Light Despite 'Risks'", *The Guardian*, 8 September 2015.
5 Smith Stegen, "Redrawing the Geopolitical Map", 89.
6 Lilliestam and Ellenbeck, "Energy Security and Renewable Electricity Trade".
7 "Crimea Officials Say Ukraine Has Cut Off Power Again", *Reuters*, 30 December
   2015.
8 Yergin, *The Prize*, 620.
9 Quoted in Högselius, *Red Gas*, 67f.
10 Hecht, "The Power of Nuclear Things", 22.
11 "Oil Giant Exits Sudan", *BBC*, 31 October 2002.
12 Yergin, *The Prize*, 572f.
13 Smith Stegen, "Redrawing the Geopolitical Map", 88.
14 "A More Assertive Ukraine Returns to Russian Natural Gas", *Stratfor World-
   view*, 12 February 2018.
15 For example, Cantoni, *Oil Exploration, Diplomacy and Security in the Early
   Cold War*, Chapter 5.
16 Högselius et al., *Europe's Infrastructure Transition*, Chapter 2.
17 Högselius, "The Saudi Arabia of the Far East?"
18 Mares and Martin, "Regional Energy Integration in Latin America".
19 Klare, *Rising Powers, Shrinking Planet*, 127.
20 For this discussion see further Stent, *From Embargo to Ostpolitik*; Högselius,
   Red Gas; Cantoni, *Oil Exploration, Diplomacy and Security in the Early Cold
   War*.
21 See, for example, Öhman, "Taming Exotic Beauties".
22 Yergin, *The Prize*, 412.
23 Högselius, *Red Gas*.
24 Klare, *Rising Powers, Shrinking Planet*, 125, quoting a New York Times interview. The "Caspian" pipeline goes from Baku via Tbilisi to Ceyhan on Turkey's
   Mediterranean coast.
25 Cioc, *The Rhine*, 67.
26 Fitzmaurice, *Damming the Danube*.
27 Avango et al., "Swedish Explorers, In-situ Knowledge and Resource-based
   Business in the Age of Empire".
28 Klare, *Rising Powers, Shrinking Planet*, 1–8.
29 See Gazprom's website, www.gazprom.com.
30 Högselius and Kaijser, *När folkhemselen blev internationell*, Chapter 6.
31 Högselius, *Red Gas*, 187.

32 "Algerian LNG Cutoff (to US) not Tied to Arab Embargo", *Oil and Gas Journal*, 26 November 1973; "Algeria May Renegotiate LNG Pacts", *Oil and Gas Journal*, 17 December 1973.
33 Högselius, *Red Gas*, 207f.
34 Ibid.
35 Vikström, "Specter of Scarcity", 1f.
36 Klare, *Rising Powers, Shrinking Planet*, 123.
37 Högselius, *Red Gas*, Chapter 7.
38 Klare, *Rising Powers, Shrinking Planet*, 170.
39 For example, Klare, *Rising Powers, Shrinking Planet*, 154–156.
40 Haynes, "Al-Qaeda, Oil Dependence, and US Foreign Policy", 62.
41 Ibid., 69–70.
42 Chapman, "Gulf Oil and International Security", 82.
43 Do et al., "Terrorism, Geopolitics, and Oil Security".

Further reading

Högselius, Per. *Red Gas: Russia and the Origins of European Energy Dependence.* Basingstoke and New York: Palgrave Macmillan, 2013.
Scholten, Daniel, ed. *The Geopolitics of Renewables.* Cham: Springer, 2018.
Yergin, Daniel. *The Prize: The Epic Quest for Oil, Money and Power.* London: Simon and Schuster, 1991.

Bibliography

Avango, Dag, Per Högselius and David Nilsson. "Swedish Explorers, In-situ Knowledge and Resource-based Business in the Age of Empire". *Scandinavian Journal of History* 43, 3 (2018): 324–347.
Cantoni, Roberto. *Oil Exploration, Diplomacy and Security in the Early Cold War: The Enemy Underground.* Abingdon: Routledge, 2017.
Chapman, Duane. "Gulf Oil and International Security: Can the World's Only Superpower Keep the Oil Flowing?" In *Energy Security and Global Politics: The Militarization of Resource Management*, edited by Daniel Moran and James A. Russell, 75–94. London and New York: Routledge, 2009.
Cioc, Mark. *The Rhine: An Eco-biography, 1815–2000.* Seattle: University of Washington Press, 2003.
Do, Quy-Toan, Jacob N. Shapiro et al. "Terrorism, Geopolitics, and Oil Security: Using Remote Sensing to Estimate Oil Production of the Islamic State". *Energy Research & Social Science*, forthcoming.
Duffield, John S. *Fuels Paradise: Energy Security in Europe, Japan, and the United States.* Baltimore, MD: Johns Hopkins University Press, 2015.
Fitzmaurice, John. *Damming the Danube: Gabcikovo/Nagymaros and Post-Communist Politics in Europe.* Boulder, CO: Westview Press, 1995.
Haynes, Peter. "Al-Qaeda, Oil Dependence, and US Foreign Policy". In *Energy Security and Global Politics: The Militarization of Resource Management*, edited by Daniel Moran and James A. Russell, 62–74. London and New York: Routledge, 2009.
Hecht, Gabrielle. "The Power of Nuclear Things". *Technology & Culture* 51, 1 (2010): 1–30.

Högselius, Per. *Red Gas: Russia and the Origins of European Energy Dependence.* Basingstoke and New York: Palgrave Macmillan, 2013.

IAEA. *IAEA Tools and Methodologies for Energy System Planning and Nuclear Energy System Assessments.* Vienna: IAEA, 2009.

Klare, Michael T. *Rising Powers, Shrinking Planet: The New Geopolitics of Energy.* New York: Metropolitan Books, 2008.

Lilliestam, Johan, and Saskia Ellenbeck. "Energy Security and Renewable Electricity Trade – Will Desertec Make Europe Vulnerable to the 'Energy Weapon?'". *Energy Policy* 39 (2011): 3380–3391.

Mares, David, and Jeremy Martin. "Regional Energy Integration in Latin America: Lessons from Chile's Experience with Natural Gas". *Third World Quarterly* 33, 1 (2012): 55–70.

Öhman, Maybritt. "Taming Exotic Beauties: Swedish Hydropower Constructions in Tanzania in the Era of Development Assistance, 1960s–1990s". PhD thesis, KTH Royal Institute of Technology, 2007.

Schrattenholzer, Leo, Asami Miketa, Keywan Riahi and Richard Alexander Roehrl. *Achieving a Sustainable Global Energy System: Identifying Possibilities Using Long-term Energy Scenarios.* Cheltenham and Northampton, MA: Edward Elgar/IIASA, 2004.

Smith Stegen, Karen. "Redrawing the Geopolitical Map: International Relations and Renewable Energies". In *The Geopolitics of Renewables*, edited by Daniel Scholten, 75–96. Cham: Springer, 2018.

Stent, Angela. *From Embargo to Ostpolitik: The Political Economy of West German-Soviet Relations, 1955–1980.* Cambridge: Cambridge University Press, 1981.

Storm van Leeuwen, Jan Willem. "Nuclear Power: The Energy Balance." www.stormsmith.nl (accessed 25 February 2012).

Vikström, Hanna. "The Specter of Scarcity: Metal Shortages in Historical Perspective, 1870–2015". PhD thesis, KTH Royal Institute of Technology, 2017.

Yergin, Daniel. *The Prize: The Epic Quest for Oil, Money and Power.* London: Simon and Schuster, 1991.

7   Energy transnationalism

The Saint-Simonian imperative

The idea that energy can be used as a foreign policy tool is usually discussed in the context of sinister threats, hawkish power struggles, violence and evil. From this perspective, energy's political utility is that it can force actors in another country to do something they would not have done on a voluntary basis. Energy then becomes a metaphorical "weapon", and the international energy arena a battle field. This is the "hard" use of energy in geopolitical affairs.

But actors also use energy as a "soft" foreign policy tool, mobilizing it for the purpose of strengthening international cooperation, fostering global political stability and economic prosperity. This phenomenon, which may be referred to as "energy transnationalism", draws on a well-established philosophical tradition. In connection with the Congress of Vienna (1814–1815), at which a new European peace order was negotiated in the aftermath of the Revolutionary and Napoleonic Wars, the French philosopher Claude Henri de Saint-Simon pointed to transnational infrastructures as important means for creating interdependencies between countries, in a way that would make future wars virtually impossible. This powerful vision has since reappeared in many forms. Today, the Saint-Simonian imperative – the idea that energy, natural resources and infrastructures can and must be mobilized to forge a peaceful world order, based on political unity and harmony – is omnipresent.[1]

As we will see in this chapter, there has been intense interaction between "hard" and "soft" uses of energy as a foreign policy tool, between energy as a "weapon" and the Saint-Simonian imperative. In fact, what from one angle appears to be an energy weapon might, from another angle, instead seem like an attempt to foster international cooperation. For example, are Russia's exports of natural gas to Western Europe a case of "hard" energy geopolitics, with the latent threat of politically motivated supply disruptions as the main feature? Or does the intricate pipeline infrastructure that materially connects East and West – and which Russian and European engineers built in close cooperation with each other – rather reflect a "soft"

significance of energy in the geopolitical realm? And should energy cooperation within the European Union (EU) be interpreted as a case of Saint-
Simonian transnationalism or rather as a hard European "weapon" in the
old world's struggle against new, rising powers? This is not clear at all.
Hence, we need to be careful when interpreting various attempts to respond
to the Saint-Simonian imperative.

The origins of European electricity transnationalism

Europe is the most fascinating region when it comes to uses of energy as
a "soft" or "positive" foreign policy tool. In the nineteenth century, the
Saint-Simonian idea of large-scale infrastructures as powerful vehicles for
international cooperation mainly targeted waterways and railways. In the
twentieth century, energy increasingly came to the fore, whereby electricity
grids attracted particular attention. Electrical transnationalism began in
earnest during the interwar era, which coincided with a technical debate in
which experts argued that electricity systems must scale up. In Britain, for
example, wartime efforts had demonstrated the fuel-saving advantages of
"power pools", in which many power plants jointly supplied a large area.
This was technically possible and economically feasible if all generators and
electro-technical machines operated "synchronously", that is, in tune at
one frequency. The key technology in this context, alternating current (AC)
transmission, had seen its breakthrough before the war. In the course of the
1920s, almost all European governments, which were now very eager to
take control over system-building, identified the striking economic benefits
of power pools as a legitimate way to strengthen their influence, envision-
ing state-wide electricity systems.

The leading system-building roles that states proposed for themselves
were challenged, however, by a cluster of actors who stressed the need for
international connections rather than national autarky. Taking inspiration from the internationalist spirit of the pan-European movement and
the League of Nations, they identified transboundary electricity systems
as a tool well-suited both to strengthen the European economy and, in
the Saint-Simonian tradition, to boost international solidarity and securing
peace. A few cross-border connections had already been built before World
War I, but they had been of only local importance. The new proposals were
much more ambitious. Especially around 1930, ideas accumulated about
turning the continent's unequal energy geography from a problem into an
opportunity, mobilizing it for the benefit of Europe as a whole. The result was a number of grand visions, featuring proposals for pan-European
power pools and transmission grids that would stretch from the British Isles
or the North Sea coast to Russia and from Scandinavia to Portugal.

The pan-European electrical enthusiasts sought to circumvent political
borders. Instead of an Electrical Europe dominated by clearly defined national networks, as imagined by state planners, they had a single European

"supergrid" – as we would call it nowadays – in mind. Only such a grid, they argued, could rationally exploit Europe's unequally distributed hydropower and coal resources. Some, notably the Germans Ernst Schönholzer and Oskar Oliven, took inspiration from corresponding pan-European network schemes in transport and communications, referring to long-distance transmission lines as "international power highways".[2] The most imaginative pan-European electrification plan was Herman Sörgel's "Atlantropa", whose cornerstone was a giant hydropower dam across the Strait of Gibraltar. Atlantropa would – physically and politically – bring Europe and Africa together in what was seen as an intensifying global struggle with two other "A's", Asia and America.[3]

None of these visions materialized. Electrical pan-Europeanism quickly dwindled in the wake of the Great Depression and a new wave of economic nationalism in the 1930s. The prospects for financing a top-down multinational power pool plummeted.[4] During World War II, Hitler's engineers sought to re-animate the idea of a pan-European electricity grid – under Nazi control – but failed to realize more than a handful of relatively minor projects.[5]

But Europe's Saint-Simonian electricity visionaries did not give up. Even as Europe was divided into a capitalist and a communist camp in the postwar era, separated by an Iron Curtain, they continued to argue in favour of electrical transnationalism. The Director General of the United Nations Economic Commission for Europe (UNECE), Gunnar Myrdal, wished to draw up a European electrification programme which would "take account of geological factors irrespective of political frontiers". The underlying argument was that energy cooperation might even serve to overcome Europe's political division into East and West. While regarding the interwar pan-European supergrid proposal as "a somewhat utopian scheme which is not economically justified", the Chairman of UNECE's Committee for Electric Power, Pierre Smits, thought that policymakers would have to learn to "think in European terms". Among other things, this translated into large-scale transfers of hydropower to Western Europe from Norway, Austria and Yugoslavia and of thermal power from Poland and Czechoslovakia.[6]

Such views were, geopolitically speaking, too radical for the two superpowers, the United States and the Soviet Union. While recognizing the benefits of international cooperation, they advocated the formation of two distinct European electricity grids – one in the West and one in the East. While each might involve strong international connections, under no circumstances should the two be allowed to become interdependent. Nature-based synergies across the Iron Curtain were thus not to be exploited; economics and efficiency were to be sacrificed for the sake of political, military and ideological considerations.

In the subsequent period, European electrification thus proceeded in accordance with a vision of transnational integration, which, at the same time, was shaped by the mutual antagonism between Washington and Moscow. Within the West, US Marshall planners and, in particular, the European

Abb. 40. **Das Raubtier „Mensch". Europa ist ein großer Käfig mit Einzelzellen.**

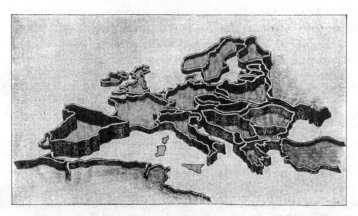

Wer es einer bloßen schönen Idee zuliebe wagen würde, seinen Käfig zu
öffnen, wäre die Beute der anderen.

Abb. 41. **Statt trennender Mauern: bindende Leitungen!**

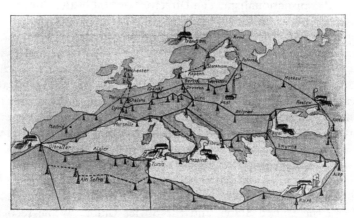

Nur eine gemeinsame, gleichzeitige Verkettung durch ein Groß=Kraftnetz
schafft eine Europa=Union.

Figure 7.1 The Atlantropa vision. Pan-European visionaries identified electricity
systems as tools for strengthening European competitiveness, boosting
transboundary solidarity and securing peace. This pair of maps shows
Herman Sörgel's radical Atlantropa vision. The perceived connection
between electricity grids and Europe's political future is here made explicit through the argument that "only a joint, simultaneous chaining
through a high-voltage grid creates a European Union". Source: Originally published in Sörgel, *Die drei großen "A"*, 91.

Recovery Programme's leading electricity advisor, Walker Cisler, pushed for a tightly integrated Western European system. The degree of integration within capitalist Europe was to be as high as possible, and national interests relegated to a subordinate role in this context. In order to maximize overall Western European benefits, projects should be "selected without regard to national frontiers". The difference, compared to UNECE's competing vision, was that Eastern Europe must be excluded.[7]

Cisler's proposal, however, clearly undercut not only the UNECE's interests but also those of Europe's regional and national power companies. During and after the war, governments throughout the continent had seized or strengthened control of the electricity systems in their respective countries, and they were by no means willing to give away that power to the prospective operator of a Western European supergrid. In the end, governments and utilities succeeded in diverting most Marshall Aid for electricity (re)construction to distinctly national projects, with cross-border links playing a subordinate role, subject to bilateral negotiations between power companies in different countries.

As a result, the triumphant model for integrating Western Europe electrically centred on distinct national or subnational systems that were interlinked with each other only on the margin. But power companies were still eager to cooperate with each other. While all countries wished to retain self-sufficiency in electricity supply, engineers identified cross-border connections as a vehicle for improving grid stability and economizing on reserve capacity. Electricity nationalism and transnationalism hence came to coexist – in a way that has come to be regarded, in hindsight, as highly successful. In 1951, company representatives from six NATO countries – Belgium, France, Italy, Luxembourg, the Netherlands and West Germany – plus Austria and Switzerland, established a joint non-governmental organization to coordinate the building and operation of cross-border power links. It was dubbed the Union for the Coordination of Production and Transmission of Electricity (UCPTE). In 1958, the power companies started synchronous operation of the entire UCPTE supply area. Although this area covered only a very small portion of the territory previously imagined as a pan-European electricity space, UCPTE presented itself as the main "European" grid.[8]

Northern Europe's electricity companies were invited to join UCPTE's power pool. But despite the fact that their region consisted of capitalist nations that were also recipients of Marshall Aid, they opted not to accept. Instead, inspired by the post-war surge in Nordic cooperation, they created a Nordic counterpart to UCPTE in 1963. The resulting divide between Northern and Western Europe originated in the technical difficulties and high costs of laying undersea synchronous links. Since the main barrier was between western and eastern Denmark, western Denmark did not join the Nordic electricity pool. Instead, it linked up with UCPTE. A number of asynchronous connections in the form of High-Voltage Direct Current

(HVDC) links, laid on the bottom of the sea, eventually enabled a certain degree of interaction between the Nordic and the Western European blocs.[9]

In southern Europe, Spain and Portugal, because of their non-democratic regimes, were not invited to join UCPTE. Instead, the main Iberian electricity companies developed their own partnership, building connections with each other. Moreover, in 1962 Spanish, Portuguese, and French power sector representatives introduced a "Franco-Iberian" electricity union, modelled on UCPTE. Two years later, a similar organization was established to enable cooperation between Austria, Italy and communist Yugoslavia. Both Austria and France thus participated in two regional electricity organizations.[10]

In communist Eastern Europe, the Soviet ambition to create a supergrid was more successful than the US-led attempt in Western Europe. The vast oil-shale reserves of the new Estonian Soviet Socialist Republic, which had catered to Nazi needs for oil in the interwar years, acquired a new function as a set of large shale-fuelled power plants came online in the 1950s. Transmission lines linked these thermal plants with Latvian and Karelian hydropower as well as with thermal capacities in Lithuania, Belarus and Kaliningrad (the former German city of Königsberg). By the mid-1960s, a "North Western Ring" had been created, the topography of which no longer provided any clues about pre-war political borders. The supergrid's control centre was in Latvia, which hosted the main balancing hydropower capacities, whereas the Estonian thermal plants, whose output far exceeded Estonia's own needs, provided the base load. Leningrad was the system's most important consumption centre.[11]

Several analogous power pools materialized further east and south throughout the Soviet Union. In 1967, Brezhnev's engineers, under the command of the powerful Soviet Minister of Electrification, Piotr Neporozhny, created a Central Dispatch Centre in Moscow, through which the different pools were synchronously linked to each other. They also built a set of new high-voltage transmission lines, most of which disregarded the boundaries between power pools in a way that would have been unthinkable in Western Europe. This allowed for massive "exports" of electricity from one pool to another, with deliveries towards Moscow playing a particularly important role.[12]

At the same time, the Soviet satellite states in Central and Eastern Europe struggled to expand electricity supply, identifying it as a crucial factor in developing heavy industry. Taking inspiration from UCPTE, national system-builders proposed forming a Central and Eastern European power pool. Party leaders enthusiastically supported the idea of this electrical communist brotherhood, while central planners viewed it as part of broader efforts to unify the economic plans of individual states. Cooperation began in 1956, when COMECON set up a commission for electric power exchange and utilization of the Danube's hydropower potential. Three years later, this body became the Standing Commission on Electric Power, expected to unite the grids of Bulgaria, Romania, Hungary, Czechoslovakia, Poland, East Germany and parts of the Soviet Union.

A previously constructed transmission line between Hungary and Czechoslovakia served as the basis for this effort. In 1960, Poland and East Germany linked their grids with Czechoslovakia, and indirectly with Hungary's network. In 1962, then, the Hungarian system merged with the network of what on the interwar European map had been eastern Poland. Upon Soviet annexation in 1944, this region had become part of Soviet Ukraine, but its electricity system had not yet been integrated with the rest of the USSR. In electrical terms, the former eastern Polish territories were thus hardly a Soviet region; its interwar political legacy remained electrically visible. Romania and Bulgaria joined the COMECON system in 1963 and 1966, respectively. Prague was the heart of the COMECON integrated system, as it hosted the central dispatch centre, whence orders could be sent to power plants and grid operators in the other communist countries.[13]

After the collapse of the Soviet Union, European electricity transnationalism underwent two important transformations. First, most of the former communist countries in Central and Eastern Europe (but not the former Soviet republics) delinked themselves from the Soviet Union and, instead, synchronized their grids with UCPTE. UCPTE hence expanded eastwards while the former grid that had connected Ulan Bator with East Berlin shrank. Secondly, the dominant role of UCPTE and the other Western European power pools was challenged by agencies of the EU, which since the 1990s has sought an increasingly powerful role in shaping European energy transnationalism. The EU Commission, in particular, argued that greater all-European coordination in energy was necessary and that the Commission would be the most natural actor to lead that process. Among other things, it pointed to new challenges such as market liberalization and new risks linked to terrorism and cyberattacks. UCPTE and the other meso-regional organizations that had existed for half a century argued that a stronger EU role would not strengthen but, on the contrary, weaken security of supply. They would not voluntarily cede the substantial power they had accumulated in European electrical affairs. Gradually, however, the organizations changed their strategy, initiating a dialogue with the Commission in which they sought to influence rather than prevent the latter's plans and reforms. Eventually, the sector organizations also followed the European Commission's suggestion that they merge into an EU-wide European Network of Transmission System Operators for Electricity (ENTSO-E). Accordingly, the old meso-regional electricity organizations were dissolved in 2009. A similar body, the European Network of Transmission System Operators for Gas (ENTSOG), was formed in natural gas. Today the EU's central role in leading European energy transnationalism is more or less taken for granted.

Appropriation of the European model

Not everyone has interpreted the European experience in using electricity to foster transnational cooperation and international political harmony as a success story. Yet it has provided ample inspiration for other regions in the world whose foreign policy actors have been seeking suitable tools to improve transnational cooperation. Take the visions of African economic – and potentially political – integration, which started to be elaborated in parallel with decolonization in the 1950s and 1960s. While it proved difficult to establish any consensus on the ultimate goal of intra-African integration, and governments strongly disagreed with each other on key issues, most did agree that infrastructural integration was something that would benefit the continent. For example, the Organisation of African Unity (OAU), which was established in 1963, early on created a Pan-African Telecommunications Union, a Pan-African Postal Union, a Union of African Railways and other infrastructural agencies. Then, following the end of the Cold War and the dissolution of the apartheid regime in South Africa, electricity increasingly came to the fore as a tool for enhancing African transnationalism. Both continent-wide electrical integration and more regionalized approaches emerged as visions in this context, the latter championed by (often rival) regional actors with acronyms such as ECOWAS, COMESA, EAC and SADC. The most ambitious plan so far has been the construction of a trans-continental electricity grid that would interconnect Cairo with Cape Town, Mombasa with Lagos. Its proponents regard it as "the ultimate solution to Africa's endemic energy problems and hallmark of African cooperation and development". Most of the efforts so far, however, have targeted the creation of a number of regional power pools: the Southern African Power Pool (SAPP, launched in 1995), the West African Power Pool (WAPP, 2000), the Central African Power Pool (PEAC, 2003) and the East African Power Pool (EAPP, 2005). Pan-African organizations such as the African Union, which was set up in 2001 to replace OAU, envisage that these pools will subsequently link up with each other – and with North African and Middle Eastern grids.

Colonial-era initiatives to exploit the Zambezi's hydropower potential for the benefit of a wider southern African region served as the point of departure and basis for the first of these power pools, SAPP. As in the European case, cooperation was initiated largely by engineers and electricity companies, who saw potential benefits in terms of optimizing the wider region's energy resources and of the prospect for the involved parties helping each other out in the case of technical disturbances. However, the Southern African Development Community (SADC) – which comprised 12 member states from South Africa in the south to Tanzania and the Democratic Republic of the Congo in the north – also became enthusiastic about the project and decided to support it because the member states' governments saw the project as a tool for international regional cooperation more generally. SAPP, the governments imagined, would be a showcase project that could inspire further integrative efforts in non-energy realms.[14]

Another region that has tried to build on the European model is Central America, where the electric utilities of six nations – with the enthusiastic backing of their governments – cooperate in the Central American Electrical Interconnection System (SIEPAC). The Spanish government and the Inter-American Development Bank (IADB) also support the project. SIEPAC started to be discussed in 1987, and in 1995 the involved actors agreed on an implementation plan. The real breakthrough came after the participating states signed the Framework Treaty for the Central American Electricity Market in 1998. The key component of the project, a grand 1,790-km high-voltage transmission line interconnecting Panama in the south with Guatemala in the north, was completed in 2013. A link was also constructed between the SIEPAC system and the Mexican grid, and another link is planned to tie up SIEPAC with Colombia's electricity system. SIEPAC is widely hailed as a showcase of international cooperation among the small Central American countries and more generally as a symbol of modernity, prosperity and peace – of great importance not least in view of the extreme levels of violence and hardship that has plagued large parts of Central America in recent decades.[15]

Coal, oil, gas: fossil transnationalism

Europe led the way not only in electricity but also in fossil fuel transnationalism. It was here that the first large-scale international coal market emerged in the nineteenth century, with Britain as the leading exporter and most other nations as importers of this increasingly strategic commodity. Political actors early on discovered the potential *risks* of the resulting dependencies. Subsequently, they also spotted political *opportunities*. World War I offered the first occasion for the Europeans – and the Americans – to practice international cooperation in coal. But the real breakthrough came after World War II, through the creation of the European Coal and Steel Community (ECSC). The background was the looming post-war shortages of energy. The perceived scarcity produced fears of growing competition between nation-states for coal and other key industrial commodities. In this situation, the victorious Western powers wished to make sure that German coal, following Hitler's defeat, was made available to other countries. They did not wish to see a return to the protectionist trends of the 1930s. France's foreign minister Robert Schuman, who came up with the idea of the ECSC in May 1950, proposed to deal with this spectre by setting up a common Western European market for coal and steel. However, Schuman was not interested merely in issues relating to commodities and industrial development. He anticipated that the common market would have far-reaching political effects in terms of promoting stability and reduce the risk of a new war in Europe. In the Saint-Simonian tradition, he famously argued that a European coal and steel community would make war "not merely unthinkable, but materially impossible".[16]

These ideas were further developed by the plan's chief architect, Jean Monnet, and by scholars of different breeds. A key idea was that of supranationalism, that is, the transfer of (some) political power to the "European" level. In this context, some political visionaries regarded the ECSC as a stepping stone to something much larger. Coal and steel were just the beginning. These two commodities would serve as levers of European unity in a wider sense. A successful ECSC would demonstrate that supranationalism was a fruitful approach, in a way that was expected to generate enthusiasm for unifying Europe not only in coal and steel but in a range of other areas as well.

The common market for coal and steel was officially opened in February 1953. In the end, however, the ECSC's importance for Europe's energy (and steel) supply was rather marginal. The reason was simple: the problem the community had been designed to solve was already being resolved in other ways. More precisely, the perceived post-war scarcity of coal was eliminated through a combination of market forces and technological advances. France's interest in joining the ECSC, for example, had been linked primarily to the prospects for accessing German coking coal. But in the course of the 1950s, improvements in transport technologies made it economically feasible to import abundant American coking coal.[17] Subsequently, coal started arriving from elsewhere in the world as well. The problem of coal under-production in Europe turned into a problem of over-production, and through the rest of the twentieth century the European coal industry's main challenge was how to dismantle itself in a controlled way as the continent switched from internal to external coal supply sources. At the same time, the strategic significance of coal as a fuel rapidly decreased as Europe turned to oil as its primary fuel.

Even so, the *political* significance of the European and Coal and Steel Community is undeniable. Coal and steel did become the stepping stones for building something bigger, and in 1958 the European Economic Community (EEC) was set up explicitly as a continuation and a widening of the ECSC experiment. Today every high-school and university student of European integration has to learn that the economic and political integration efforts in Western Europe started with the ECSC. Coal and steel transnationalism is seen to have fostered a politically united Europe.

In oil, there was never much internal European production worth mentioning, with the notable exception of Romania. Likewise, there never emerged any counterpart, in the oil industry, to the ECSC. Yet state and private actors in a range of European countries did initiate cooperation with each other on a grand scale, culminating in the construction of a number of transcontinental oil pipelines in the 1960s and 1970s.

The starting point for this oil transnationalism was the need for oil refining. Starting in the interwar era, European governments strongly advocated that the oil used in their countries be refined domestically rather than in the country of crude extraction. Some countries even created state-owned refining companies, which were then typically integrated with companies that were looking for oil in colonial regions. Italy's Agip is a case in point here; Spain's CAMPSA, a state import monopoly founded in 1927, another. Overall, oil refining became very much a national affair. The refineries were usually built in the immediate proximity of large seaports. At these locations, crude oil from overseas was refined and then shipped to industrial users, filling stations, airports and so on by way of tanker, rail or truck.

But oil nationalism soon gave way to a more transnational approach. When oil use started to grow exponentially in the 1950s, it soon became infeasible to rely on road and rail transport of refined oil products from the ports to major inland consumption hotspots. The stress on the European railway network, in particular, became an enormous problem. Instead, oil companies along with other actors, such as petrochemical companies, industrial groups and regional governments, joined forces to construct very large, and above all very long, oil pipelines from the seaports to major industrial centres in continental Europe, such as Ludwigshafen and Ingolstadt in Germany or Schwechat in the outskirts of Vienna. The most important oil pipelines in Western Europe were the ones from Marseilles to Baden-Württemberg, from Rotterdam to the Ruhr, from Genoa to Ingolstadt and from Trieste to Ingolstadt and Vienna.[18]

Bavaria and its political leader Otto Schedl played a particularly interesting role here, as already touched upon in Chapter 3. Schedl actively pursued a strategy that aimed to make Bavaria economically independent from northern Germany. Imported oil from North Africa and the Middle East became his tool in this struggle. However, in order to access foreign oil without relying on northern Germany, the Bavarians needed to forge cooperation not only with exporters of crude oil but also with actors in Italy, Austria and Switzerland, which emerged as transit countries. The construction of oil pipelines from Italian seaports to Bavaria, taking the route across the Alps, must be regarded as one of the most impressive feats in European oil history. But there are several other examples of landlocked regions that followed the same strategy, with actors forging cooperation with neighbouring European countries so as to enable imports of crude oil from far away. Ludwigshafen, home to Germany's powerful BASF chemical company, accessed Middle Eastern oil not so much by way of Rotterdam or Hamburg, but rather through a cooperative venture with France. In Eastern Europe, Czechoslovakia, Hungary, Poland and East Germany forged similar bonds, albeit under a totally different political regime, with the Soviet Union, whose immense oil riches they accessed through the Druzhba ("Friendship") pipeline system. The point here is that oil pipeline construction in both Western and Eastern Europe became the basis for far-reaching cooperation between different European countries. Or, somewhat differently expressed, oil pipeline construction became the basis for a dynamic

Figure 7.2 The control panel of the Druzhba system at Almetyevsk, 1971. In the
post-war decades, oil replaced coal as the Soviet Union's lead energy
source. The Druzhba ("Friendship") oil pipeline system – the world's
largest – fed oil from the Tatarstan and Samara oil fields to refineries
and chemical complexes in most Comecon member states. The Soviet
leadership hoped that grand energy projects like this would stimulate
political cohesion among the communist countries. Photo credit:
Science Photo Library.

European oil transnationalism. To what extent this had any influence on
European political integration efforts, however, is unclear.[19]

European natural gas transnationalism has also been very prominent –
and contested. It took much longer for natural gas than for coal and oil to
become traded internationally, mainly because it was so much more difficult to transport gas over long distances. Even today, there is no truly globalized market for natural gas. The breakthrough for natural gas as a fuel
in Europe is linked directly, ironic as it may seem these days, to attempts to
strengthen European energy independence. Natural gas was spearheaded in
Europe by a number of countries and regions that did not possess any large
domestic coal deposits, especially France, Italy, Austria and also Bavaria.
All of these promoted natural gas as a way to overcome coal import dependence. Their enthusiasm for natural gas was initially linked to quite
substantial domestic gas finds. But when they ambitiously set out to exploit
these deposits, natural gas soon became so popular among users that overall demand very quickly outpaced the levels that local sources were able
to meet. In this situation, the gas companies faced a delicate and strategic choice: either they would have to disappoint their gas users, informing

them that because, unfortunately, there was no gas left, they would have to switch to coal or oil. Or they would have to try and meet growing demand by seeking access to foreign supplies. Virtually all gas companies in Europe decided to pursue the second alternative.[20]

But from where could the gas be sourced? Who could possibly supply Europe with more natural gas to compensate for depleted local supplies? The first large-scale project to be seriously discussed here aimed at moving gas from North Africa across the Mediterranean into continental Europe. This possibility became immensely popular not only in France, which had discovered a vast natural gas field in the Sahara desert, but also in Bavaria, Italy, Spain, Austria and even Britain. Plans were developed for the construction of very long-distance pipelines that would bring the gas across to southern Europe and further north. In essence, this represented a European attempt to exploit the natural resources of its colonial periphery. But what may thus appear as a colonial vision was also a vision of European energy transnationalism, because it rested on the idea of far-reaching cooperation in pipeline construction among several European countries.

Precisely at this time, however, further significant natural gas discoveries were made in Europe itself: in the Netherlands. The Groningen natural gas field there turned out to be so immense that it seemed impossible for the Netherlands itself to make use of all the gas. For this reason, the Dutch concluded agreements with Belgium, France and Germany for large-scale gas exports. This cooperation became another cornerstone of European gas transnationalism.[21]

Further and even more substantial gas fields were being discovered in the Russian colonial periphery: in northwestern Siberia. The gas discoveries there in the early 1960s made headlines in Western media, and several Western European gas companies started to consider the possibility of importing Soviet natural gas. The Soviet Union's political and industrial leaders liked the idea. And so by the 1970s, Western Europe was sourcing most of its gas from the Sahara desert and the Siberian tundra. Together with the gas from the Netherlands, these colonial gas streams made the European gas market very competitive. Despite the fact that Europe had very little indigenous gas, save the Groningen field and the smaller French field at Lacq, the competitive situation meant that gas importers were able to source large volumes of natural gas at a very favourable price. Moreover, the gas transnationalism that was already becoming evident in the early Algerian and Dutch visions became increasingly pronounced, because long-distance pipelines from Siberia, from the Sahara and later from other gas-exporting regions as well, forced entire groups of Western European countries and regions to cooperate with each other in the construction of these huge pipeline systems. Foreign policymakers generally encouraged and sanctioned the huge pipeline projects, seeing political opportunities in the transnationalization of natural gas. Willy Brandt, in particular, West Germany's foreign minister and subsequently chancellor, saw the emerging

Soviet-German gas trade as a way to generate trust in the deeply problematic relations between Germany and the Soviet Union.[22]

Europe is not the only region where oil and gas transnationalism has featured prominently. For example, a dense network of oil and gas pipelines stretches across the long US-Canadian border, with substantial flows in both directions. This transnational system has forced US and Canadian private and public actors to cooperate with each other, in a way that has served well to strengthen the ties between the two countries more generally. Today, Canada is widely regarded as a guarantor of US energy security. Oil and, increasingly, natural gas, also play an important role in US-Mexican relations.

South America is another region that has seen ambitious attempts to combine transnational energy system-building, especially in natural gas, with political regional integration. As in Europe and North America, this development has been driven by bilateral efforts rather than orchestrated by any pan-regional actor like the ECSC in coal or the IAEA in nuclear issues. But a range of international organizations, such as the IADB and the Organización Latinoamericana de Energía (OLADE), have been drawn into the various projects both for funding and for technical consultation. The first visions of South American gas integration originated back in the 1950s, with the first transnational pipeline, connecting Bolivia with Argentina, going operational in 1972. Over the next 20 years, little happened. In the 1990s, however, the idea of regional infrastructural integration as a powerful lever of societal development again became fashionable, and several countries identified natural gas as a vehicle both for strengthening their own national development and for improving international political relations in the region.[23]

Chile, for example, under the Concertación government, initiated negotiations with both Bolivia and Argentina for gas imports in 1991. Five years later, at the 1996 OLADE annual meeting, President Gonzalo Sánchez de Lozada of Bolivia, which had sizeable natural gas reserves, announced that he intended to turn his country into "the principal conduit for natural gas distribution in the Southern Cone". Further north, Hugo Chávez, Venezuela's controversial socialist president from 1999 to 2013, proposed to make his country's gas resources available to neighbouring Colombia and, above all, to the potentially vast gas markets of Brazil and Argentina. Chávez was very explicit about the connection between energy cooperation and political transnationalism, regarding oil and gas integration in Latin America as a key building block for an anti-US "socialism of the twenty-first century". In addition to natural gas, Chávez envisioned a later phase that entailed building an oil pipeline from the Orinoco River's oil fields in Venezuela to the port of Tumaco on Colombia's Pacific Coast, to enable the radical restructuring of Venezuelan oil exports away from the country's traditional markets in the United States to the booming economies of Asia.[24]

In terms of actual construction, Argentina and Chile built a small pipeline interconnecting their southern regions in 1996. A larger project was launched a year later, connecting central Argentina with the Santiago area in Chile. Other pipelines were built between the two countries' northern regions. Shortly thereafter, in 1999, the most ambitious of all South American pipelines was commissioned, a 3,150-km project linking Bolivia's gas fields with users in southeastern Brazil. This was followed in 2002 by the inauguration of the Cruz del Sur pipeline from Argentina to Uruguay's Colonia del Sacramento and Montevideo. A few other projects were also successfully completed. Overall, by the early twenty-first century, South America seemed to be well on its way to emulating the European success in creating a tightly integrated regional grid.

Very soon, however, the spirit of transnationalism faltered. Shortly after the Bolivia-Brazil pipeline was opened, Brazil faced an oversupply of domestically produced gas and found that it did not really need the imported gas, at least not for the time being. The Brazilians sought to use their domestic gas abundance as a bargaining chip to renegotiate the gas price with the Bolivians. This led to radical nationalist backlashes in Bolivia, ultimately resulting in President Gonzalo Sánchez de Lozada being overthrown by riots in 2003, precisely "for trying to make Bolivia the hub of regional gas integration".[25] In the subsequent period, Bolivia failed to attract investments to its gas industry, so that by 2007 the country faced a gas shortage. It could not live up to all its export commitments and decided to "reduce its supply to Argentina by over fifty percent in order to fulfil its contracts with Brazil". At the same time, Argentina itself, plagued by a severe economic crisis, was unable to live up to its contractual obligations *vis-à-vis* Chile and Uruguay. For a decade, the gas stopped flowing between Argentina and Chile.

In the next phase, judging that the prospects for regional supplies to cover gas demand in South America were too bleak, gas companies in Chile, Brazil and Argentina set out to construct LNG regasification facilities on their coasts, eyeing gas imports from elsewhere as the way forward.[26] It is only recently that South American gas transnationalism has regained some momentum, whereby imports from overseas, not unlike in Europe, are seen to stimulate regional integration; Argentina, for example, is now able to import cheap US shale gas by way of transit through Chile – using the pipelines that were originally constructed for flows in the opposite direction. As for Venezuela's export potential, however, the disastrous and deeply tragic political and economic development that has followed Chávez's death is likely to prevent any transnational initiatives for the foreseeable future.[27]

There are many other examples of oil and gas projects, especially bilateral ones, where the Saint-Simonian spirit of infrastructural cooperation as a basis for improved political relations has figured quite prominently. Israel, for example, has a long-standing cooperation with Egypt in natural gas, featuring joint pipeline construction and mutually beneficial energy trade in a way that would seem to contradict the history of difficult political relations between these two countries. The optimists believe that energy projects like this have the potential to "stabilize Israel's relations with its neighbors by creating a web of mutual interests and opening up the possibility of regional cooperation beyond the subject of natural gas". In other words, cooperation in natural gas would function as a stepping stone for building long-term political stability and avoiding future wars in the Levant.[28] China's cooperation with Japan in the 1970s and 1980s is another case in point; here the prospects for Japanese technology sales to China and Chinese oil exports to Japan motivated the two countries to put a formal end to the war fought decades earlier. Pipeline construction and oil and gas shipments are also being viewed as potentially positive vehicles for resolving difficult political conflicts in the cases of Sudan's oil relations with newly independent South Sudan, and Russia's interest in supplying both North and South Korea with natural gas.

Oil and gas resources in border regions should also be mentioned. These are often pointed to as problematic and strategically sensitive. There are many cases of conflicts over the ownership of internationally shared off-shore oil and gas fields, in particular. From a Saint-Simonian point of view, however, such cases also bear potential for the efficient resolution of historical conflicts. The cross-border natural gas field of Zwerndorf-Vysoká, for example, was seen to improve the political atmosphere between Austria and Czechoslovakia during the Cold War period.[29] More recently, the promise of large-scale oil and gas resources in the Barents Sea stimulated the Russian and Norwegian governments not only to cooperate in exploring these subsea riches but also to officially settle the sea border between the two countries, which had been subject to dispute for decades.[30]

Nuclear transnationalism

After World War II, nuclear energy emerged as a potential new energy source, and ever since, the attempts to harness it have been subject to an intricate interplay between national and transnational visions. In the beginning, there was fear: of a nuclear apocalypse, and of national security – and survival – being jeopardized by the revolutionary new weapon that the United States had developed. Washington early on anticipated that it would be impossible, in the medium term, to prevent other countries from acquiring the knowledge necessary to develop an atomic bomb. This became the main motivating force for Washington to seek international cooperation in nuclear matters. The US strategy that gradually emerged was to use the lure of nuclear fission's peaceful uses – where the US, thanks to the Manhattan Project, was far ahead of everybody else – to persuade other governments to relinquish any plans for nuclear weapons.

Already in December 1945, just four months after the destruction of Hiroshima and Nagasaki, the United States, Britain and the Soviet Union agreed that an international atomic energy commission should be created "to consider problems arising from the discovery of atomic energy and related matters". It was placed under the newly launched United Nations and was made subject to the direction of its Security Council. A year later, the commission's US representative, Bernard Baruch, proposed the creation of an International Atomic Development Authority (IADA), which would be entrusted with "managerial control or ownership of all atomic energy activities potentially dangerous to world security". This would be combined with far-reaching transnational information exchange. But many analysts considered Baruch's plan unrealistic, dangerous or simply unnecessary. Congress opposed the plan, and it never materialized. Instead, the United States entered a development path focused on secrecy in nuclear affairs and on monopolizing access to the world's uranium. The Americans joined forces with Britain and Canada, forming a joint agency that "sought to corner the market by arranging to buy all the uranium that Belgium was producing in the Congo and all that South Africa and Australia would later produce, as well as US and Canadian production". For several years, the agency "succeeded in buying almost all the uranium mined outside the Soviet Bloc".[31]

There were several actor categories in favour of greater openness in nuclear matters. Scientists and engineers wished to resume the far-reaching transnational exchange in nuclear physics and chemistry that had characterized the interwar years. American corporations that had contributed to the Manhattan Project, for their part, wished to capitalize on their expertise by selling nuclear technology to foreign countries. Their managers complained that if they were not allowed to export their nuclear products, they would be overrun by foreign competitors and lose markets. These arguments grew louder over the years as the prospects for civilian nuclear energy increased.

The turning point in the debate came with the Soviet Union's successful testing of an atomic bomb in September 1949; it came as a shock to the United States, whose experts had expected that it would take at least 20 years for the Russians to assemble their first nuclear weapon. In October 1952, Britain became the third nuclear power. In what followed, the new US President, Dwight Eisenhower, was able to convince his nation that international cooperation was the only feasible way forward. In December 1953, Eisenhower gave a famous speech at the United Nations, dubbed "Atoms for Peace", in which he proposed the establishment of what would become the International Atomic Energy Agency (IAEA). Taking inspiration from Baruch's earlier ideas, he envisaged that IAEA would have a central role not only in regulating nuclear issues worldwide but also in actually managing flows of uranium and other nuclear materials through a centralized pool or bank. That idea would turn out to be unrealistic, although even today the IAEA, which was formally established in 1957, continues to argue vigorously in favour of transnational nuclear centres for enrichment, reprocessing, nuclear-waste handling and so on.

The "Atoms for Peace" programme entailed another important component. The US offered to provide other countries with research reactors,

Figure 7.3 Atoms for Peace. US President Dwight D. Eisenhower addresses the General Assembly of the United Nations in New York on 8 December 1953. Eisenhower proposed the creation of a new UN atomic energy agency, which would receive contributions of uranium from the United States, the USSR and other countries "principally concerned" and would put this material to peaceful use. Seated on the presidential platform are (left to right) Mr. Dag Hammarskjöld, Secretary General of the UN, Madame Vijaya Lakshmi Pandit of India, President of the UN General Assembly, and Mr. Andrew Cordier, Executive Assistant to the Secretary General. Photo credit: United Nations/IAEA.

including their fuel. The Soviet Union started to do the same, and a competition between the two superpowers emerged in this field. The development was boosted by the First International Conference on the Peaceful Uses of Atomic Energy, held in Geneva in August 1955. By 1959, the US had concluded reactor agreements with no less than 42 countries, while the Soviets made progress among the communist countries, concluding agreements with 26 countries by 1968. These agreements became powerful instruments in strengthening overall political ties between the countries involved. Both the US and the Soviet research reactor offers were linked to the return of the spent nuclear fuel, so that the "back-end" of the nuclear fuel cycle, too, became internationalized. In the Soviet case, this procedure was later followed up by similar but much larger fuel supply arrangements for full-scale commercial reactors.[32]

There were several other transnational initiatives in this formative phase of the global nuclear energy system, some of which seemed to compete both with the IAEA regime and with the quest by the US and the Soviets for superpower hegemony in global nuclear issues. One was the creation of the European Atomic Energy Community (EURATOM). Its immediate source of inspiration was the 1956 Suez Crisis, which highlighted Europe's vulnerability to oil supply disruptions. Alternative sources of energy suddenly appeared more attractive. Nuclear energy enthusiasts in science and industry used the occasion to call for what they regarded as necessary increases in nuclear energy investments. Western European governments, for their part, liked the idea of a nuclear community because they saw it as a chance to further strengthen intra-European political stability and cooperation. A report published in May 1957 recommended that the six member states of the ECSC broaden their cooperation to include nuclear energy. It recommended that the member states jointly build no less than 15,000 MW of installed nuclear power by 1967. A few months later, in January 1958, they set up EURATOM and, in parallel, the EEC (the predecessor of today's EU). Washington, identifying the European moves both as a threat and as an opportunity, responded by seeking an agreement with the Europeans under which the United States would supply the enriched uranium that would be needed for the European nuclear power plants.[33]

In parallel with the launch of EURATOM, the Organization for European Economic Co-operation (OEEC, which was later to become the OECD), also became active in nuclear matters. The OEEC's initial function had been to organize European post-war recovery under Marshall Aid, and its membership was far larger than that of the six-country EEC. In December 1956, the OEEC's Council of Ministers approved the creation of a European Nuclear Energy Agency (ENEA, later to become the NEA). Under its auspices, the Europeans would launch a security control system and, more radically, build a European spent-fuel reprocessing plant (EUROCHEMIC) as a "joint undertaking".[34] The latter idea, the latest in an already long history of Saint-Simonian infrastructural initiatives, contrasted sharply with alternative, nationally oriented plans to build reprocessing plants, and attained significance as a symbol of European political unity in the emerging nuclear age. EUROCHEMIC's first facilities were built at Mol in Belgium, with construction completed in 1960. American experts were instrumental in helping the Europeans to design and construct the plant, so in fact the project furthered trans-Atlantic relations as well. ENEA further launched two experimental research reactor projects in the late 1950s.

In the 1960s, both EURATOM and ENEA seemed to lose some of their momentum, a major reason being that nuclear power programmes became increasingly nationally oriented. Attempts elsewhere in the world to emulate the European projects did not fare much better. For example, the idea of an Asiatic Atomic Energy Community (ASIATOM), as proposed by the Chairman of the Pakistan Atomic Energy Commission, came to naught.[35]

Yet the idea of mobilizing nuclear energy for political purposes lived on. In the late 1960s, for example, Britain sought to draw on its technical advances in uranium enrichment as a way to strengthen its political relations with continental Western Europe and loosen its dependence on the United States in the nuclear field. Britain had taken an early interest in joining the EEC. In the 1960s, with the British economy facing problems, this interest grew stronger. But in 1967, French President Charles de Gaulle once again vetoed Britain's application for membership. The British government then considered alternative ways to move closer to continental Western Europe, soon identifying uranium enrichment technology as a promising political tool. British scientists and engineers had made impressive advances in what they referred to as gas-centrifuge technology, a novel enrichment technology. Following the British failure to access the EEC, Britain now offered the Netherlands and West Germany access to the innovation. Germany had thus far cooperated with France in uranium enrichment issues. This cooperation seemed highly promising to the Germans, not least because France had managed to take control of large-scale African uranium resources. The British government, however, sought to convince the Germans that they could offer something better: it appeared that Britain's gas-centrifuge technology was economically superior to the gas-diffusion enrichment method that had so far dominated the market, and British mining companies also controlled substantial overseas uranium reserves. In this way, the British deliberately sought to weaken the Franco-German axis in nuclear cooperation and, by extension, forge a new political power balance in Europe. Tony Benn, UK Minister of Technology at the time, reasoned that "to isolate France with its high cost, low efficiency, nuclear technology, and secure for ourselves a commanding position in the nuclear field in the Six [EEC member states] might be a powerful political weapon". An agreement was finally reached in November 1969, paving the way for joint British-German-Dutch construction of two gas-centrifuge enrichment facilities, one in Britain and one in the Netherlands near the German border. In the 1980s, a third facility was added, this time in Germany.[36]

The most far-reaching transnationalism in nuclear energy was probably the one that took form in the Soviet Bloc. It was a strongly asymmetric transnationalism, in which the Soviet Union's leading role was unquestioned. But the degree of international exchange and cooperation was impressive. The construction of research reactors, mentioned above, was just the beginning. The Soviets helped to build up a strong nuclear engineering competence in various union republics, from Armenia to Ukraine, as well as in the satellite states of Bulgaria, Hungary, Czechoslovakia, Poland and East Germany. The research reactors were followed by full-scale commercial nuclear power plants, which were built in close cooperation among several COMECON member states. The deals for the commercial reactors included a nuclear-fuel supply regime that rested on Soviet supply of enriched uranium and fuel element fabrication, in combination with the return of all spent fuel to the Soviet Union for reprocessing.[37]

The Soviet government regarded nuclear energy as a powerful way to tie the union republics more closely to each other, as is also evident from the geography of nuclear power plants in the Soviet Union. Many of the full-scale nuclear facilities were built in inter-republican border areas, the idea being that such projects would stimulate political integration and interdependence among the constituents of the vast communist empire. Thus, for example, the Chernobyl nuclear power plant in Ukraine was situated right on the border to Belarus, whereas the Ignalina plant in Lithuania was near the border with both Russia and Belarus.[38] Yugoslavia followed this example. The Tito government decided that the country's only nuclear power plant would be built precisely on the border between Slovenia and Croatia – and that it would be jointly managed by the two republics.

Energy systems as the ghosts of empire

Energy systems may play an important role in stabilizing political relations in times of geopolitical turmoil. For example, when the British, French and other European empires were dissolved in the mid-twentieth century, giving way to a range of new, independent nations, the metropoles looked for ways to retain some degree of political and economic control over their former colonies. Energy was identified as a promising tool in this context. France, for example, lost formal control over Algeria in 1962, but at that time far-reaching efforts were already under way to exploit Algeria's vast oil and gas resources. The French and their former colonial subjects had strong incentives to continue working together on these energy projects after Algerian independence, as the new nation needed access to French technology and capital, and France was in dire need of Saharan oil and gas. The way forward was not easy, given the traumas, emotions and violence involved in the political "divorce". At times, the new independent government in Algiers sought to avoid cooperation with the French, turning, instead, to partners from Italy, Britain and elsewhere. Yet France continued to play a unique role in Algeria's foreign relations and Algerian oil and gas contributed decisively to strengthening these.

The French proceeded in similar ways in Sub-Saharan Africa. African uranium, in particular, became a matter of existential importance to France in the nuclear age. The country's colonial possessions emerged as guarantors of an independent nuclear fuel supply. The Commissariat for Atomic Energy (*Commissariat à l'Énergie Atomique,* CEA) early on sent out its geologists to map the uranium resources in Madagascar, Gabon and Niger. Like in oil, they soon faced the daunting challenge of sustaining their operations in a turbulent era of decolonization. In accordance with a treaty signed in connection with Niger's independence in 1960, Niger gave France priority access to its uranium. After years of debate and repeated interventions by Presidents Charles de Gaulle and Georges Pompidou, two French-controlled mining companies were tasked to direct the build-up of a Nigerien uranium industry. By the early 1970s, Niger's Saharan uranium resources were considered "indispensable to the French energy boom". They also ensured that the former colonies remained tied to Paris.[39]

Similar stories can be told about the other European imperial powers and their efforts to mobilize energy to retain fruitful relations with their former colonies. The era of decolonization was turbulent and disruptive, but energy offered a way forward. Critics have pointed to Western-led extraction of energy in the former colonial world as a typical case of "neo-colonialism" or "neo-imperialism".[40] They argue that the Western powers have continued to control their former overseas possessions, especially in economic terms, the only difference being that they no longer exercise any *formal political power*. It is also clear that, to the extent that these neo-colonial relations have been fruitful, they have not necessarily benefited everyone, but first and foremost a privileged political and economic elite. But regardless of whether we choose to interpret post-colonial relations as good or bad, there is no doubt about the fact that energy has played an important role in sustaining the overall political and economic relations between the former Western imperial powers and the now independent countries in the developing world.

Another empire that eventually collapsed was that of the Soviet Union. In this case, the situation was different, as the energy links between Russia and the other Soviet republics had grown very intricate over the preceding half-century. The existing systems were already mature and were characterized by a high level of momentum. As we have seen, Soviet planners had deliberately designed the systems in such a way as to maximize interdependence between different union republics. From a systems perspective, this seemed to make it logical and rational to maintain fruitful energy relations among the now independent ex-Soviet nations. The latter, however, were generally keen to achieve independence from Moscow in all fields, including energy.

The three Baltic countries, having gained full political independence in August 1991, immediately probed the prospects for delinking their electricity grids from those of Russia and Belarus. They were now politically free to do so. The only problem was that the technical stability of the existing system rested on tight, synchronous high-voltage transmission within the North Western Ring, which now comprised the Baltics plus Belarus and northwestern Russia, and that both Estonia and Lithuania were major net exporters of electricity to their eastern neighbours. The idea of delinking the Baltics thus entailed significant supply risks, in the form of potential blackouts and other grid instabilities, and economic risks in the form of lost export revenues. The latter were important for the Baltics due to the deep economic recession the republics faced in the aftermath of communism's collapse. In this situation, the three countries approached their Nordic neighbours, suggesting that the Baltic electricity grids might be connected with the powerful Nordic system, and also with Poland, which at this time was seeking integration with the Western European grid. Such interconnections, the proponents reasoned, could strengthen Baltic grid stability and also offer an alternative outlet for surplus electricity production in the region.[41]

What has become of these visions? In 2006, a decade and a half after communism's collapse, the Estonian and Finnish grid operators managed to interlink their grids through a submarine HVDC link across the Gulf of Finland. More recently, the Lithuanian grid was hooked up with its Polish counterpart and the Latvian grid with the Swedish high-voltage system – also through HVDC cables. However, since Lithuania was forced to shut down its massive nuclear power reactors at Ignalina as a precondition for its EU accession in 2007, while Estonia's overproduction of electricity from highly polluting oil-shale combustion has also decreased, the Baltics have not been able to earn any significant export revenues from electricity exports to Scandinavia or Poland. More surprisingly, nearly three decades after the Soviet Union's demise, the three Baltic countries have not delinked their networks from those of Russia and Belarus. They are still synchronously interconnected through the North Western Ring, built during Soviet times, which continues to operate much as before. In political terms, Estonia, Latvia and Lithuania have been members of the EU and NATO since 2004. In electricity, by contrast, they are still part of what used to be the Soviet Union. Hence, the Soviet Union, or at least its ghost, continues to exist in electricity.[42]

Are such lingering ex-Soviet interconnections good or bad? The most common interpretation among Western analysts is that they are bad. Their argument is that they embody problematic dependencies on Russia, and that Russia might abuse these dependencies. This discourse is most pronounced in the case of natural gas, where the Baltics continue to rely almost exclusively on Russian supplies. In the case of electricity, however, the relationship is less asymmetrical, as the Baltics are dependent on Russia to about the same extent as Russia (and Belarus) depend on the Baltics. Since the dispatch centre for the North Western Ring is located in Riga, the Latvian capital, rather than in Russia, one might even argue that the Baltic grid operators have greater leverage over their Russian counterparts than the other way round. For example, the dispatchers in Riga could conceivably disrupt cross-border electricity flows to Russia in such a way as to cause a blackout in St. Petersburg. However, this has never happened. Instead, the grid operators in Estonia, Latvia, Lithuania, Belarus and Russia cooperate on a very fruitful basis with one another. In fact, this cooperation may be interpreted as a showcase of constructive transnationalism in an age of extremely tense political relations in the region. In the Saint-Simonian spirit, electricity transnationalism in the former Soviet Union bears the potential for a friendlier political future.

Energy transnationalism in the age of renewables

What does the Saint-Simonian imperative signify in the construction of renewable energy systems? The transition to renewables has so far most commonly been associated with environmental opportunities, especially in the context of climate change. They have also, to some extent, been regarded as tools to cope with problematic dependencies on fossil fuel imports. China is the country that most vigorously pursues renewable energy investments for energy security purposes.[43] But there is a third and arguably even more visionary dimension relating to the energy transition: renewables hold the potential to strengthen international political stability, foster trustful relations between geopolitical enemies, prevent unnecessary wars and, more generally, stimulate peace and harmony in the world. Such a claim might sound naïve and perhaps counter-intuitive. Yet it is part and parcel of energy and geopolitics in the age of renewables.

Take the rise of the international biofuels trade in the early 2000s. It was accompanied by loud complaints about its negative environmental effects in the Global South and its impact on food prices in the developing world. Agricultural lobbyists in the Global North played an active role in this debate, viewing biofuel imports from countries such as Brazil as a form of unwelcome competition to European and North American biofuel feedstock production. Some visionaries, however, argued that the environmental and economic problems relating to biofuel production in the Global South could be dealt with in effective ways, while also pointing to the political opportunities of a biofuels transnationalism. Their argument was that biofuel feedstocks, undoubtedly, could be grown much more productively in tropical countries than in the cold Global North, and that, as a consequence, it would not make any economic or environmental sense for the developed countries in the North to pursue protectionist biofuels policies oriented towards domestic feedstock production. Instead, both the South and the North had a lot to gain from a large-scale international trade regime based on biofuels exports from the South to the North. For this reason, a "Biopact between North and South" should be launched. "If the United States and Europe are serious about biofuels, they must turn to the South for their supplies", stated the IEA's general director, Claude Mandil, in October 2006. John Mathews, another visionary, argued that a North-South Biopact

> would represent a practical and powerful way for the North to act to protect biodiversity and help countries in the South to prevent deforestation, as opposed to the hand wringing that passes for action at the moment. This could be a Biopact with enormous consequences for both North and South, and as such it could help to shape an international regime of peace, security and economic development for the twenty-first century.[44]

Soon afterwards, the EU Commission and the Brazilian government under President Lula da Silva used the momentum of the biofuels debate and the prospects of Brazil as a major supplier of bioethanol to the EU to forge a strategic partnership. At a much-publicized EU-Brazil Summit held in July 2007, the two sides agreed to "coordinate their stance" on a number of key global issues, ranging from democracy and peace to poverty and human rights. Initiatives were also taken to forge an association agreement between the EU and Mercosur, the South American trade bloc.[45]

Or consider the bold visions of constructing modern supergrids for efficient electricity transmission. As pointed out above, scientists, engineers and electricity companies typically frame these as solutions to the problem of intermittent production of wind and solar energy. Hawkish foreign policymakers fear them because they might generate new forms of dependencies on other countries. Proponents of localized, decentralized electricity systems hate them for their sheer size. Twenty-first-century Saint-Simonian visionaries, by contrast, love them.

In Europe, the EU Commission has for decades sought to stimulate greater trade and interdependencies between its member states. The nation-centred European electricity systems that took form in the twentieth century have worked against this logic, so that even today, the amount of electricity exchanged across borders remains marginal. The supergrid visionaries promise to change this, and thus have become very popular with the Commission. For the Commission, the purpose of projects such as the "North Sea Countries Offshore Grid Initiative", launched in December 2009 by the electric utilities and foreign ministers of nine European states and with the Commission's support, is not merely to strengthen renewable electricity supply in the EU. It is seen as a way to bring the participating nations closer to each other in a more general political sense. It is about strengthening the EU's relations with Norway – not only in electricity but at large. And following the 2016 decision by the United Kingdom to leave the EU, North Sea energy cooperation has become one of the fields in which the EU and Britain will have to retain a very close relationship.

The bold visions of electrically interconnecting Europe with North Africa and the Middle East are even more significant from a Saint-Simonian point of view. A supergrid spanning the Mediterranean, clearly, would not merely be a technical and environmental project, built for the purpose of improving electricity supply and reducing carbon emissions. It would also constitute a major political and cultural undertaking. At focus are here the relations between regions that are radically different in political, economic and cultural terms. North Africa and the Middle East used to be a European – and Turkish – colonial periphery, with all of the negative connotations this entails. Since decolonization, there has been a great deal of anti-European sentiment on the Mediterranean's southern shores, while in Europe, anti-Muslim sentiments are currently growing and there is a constant fear of Islamist terrorist attacks. Can renewable energy transnationalism help to counter these trends and improve overall Mediterranean relations? The region's supergrid visionaries believe so.

Supergrid transnationalism in Europe has subsequently inspired a range of similar initiatives elsewhere in the world, especially in Asia. Hence, Desertec, for example, inspired the East Asian Gobitec vision, which centres on solar electricity transmission from Mongolia and China to major consumption hotspots throughout East Asia. In Japan, the 2011 Fukushima nuclear disaster provided further impetus in this direction. Masayoshi Son, "the richest man in Japan" and founder of the Tokyo-based Renewable Energy Institute, in this context proposed the much-publicized "Asia Super Grid" concept. Its purpose would be not merely to improve electricity supply and combat climate change but also to create jobs and alleviate poverty. Above all, the project would "lead to closer cooperation between the countries, thus reducing national animosities that began before the Second World War and still persist".[46]

The United Nations has also been extremely enthusiastic about the supergrid visions. The UN perspective on electricity transnationalism, firmly rooted in the Saint-Simonian tradition, illustrates better than anything else the geopolitics of hope in twenty-first-century energy affairs. As such, it may conclude not only this chapter but also our book as a whole:

> The planning, design, construction, and operation of a grid interconnection between two (or more) nations requires cooperation of many different types. High-level political cooperation between countries is certainly necessary, but a potential benefit of grid interconnection is also that the project can serve as a spur to cooperation at the societal level as well. If the grid interconnection serves to provide (or enhance) a political bridge between nations, the bridge can be used to foster social exchanges in sports, education, and culture, for example, promoting understanding between societies. Similarly, enhanced trade in other commodities between countries could follow from the experience in trading electricity, bringing citizens from the interconnected societies into additional contact in the process. In addition, grid interconnection activities, such as power line construction and maintenance, or construction of new power stations, may, depending on how the contracting crews are selected, bring workers from the interconnected countries together. Working together on projects of clear mutual benefit, and working in ways that provide person-to-person contact between people of different nationalities, is an excellent method of building trust and understanding between peoples from different societies.[47]

Exercises

- Select an international (bilateral or multilateral) energy project and discuss the extent to which it appears to be driven primarily by "hard" or "soft" considerations in the geopolitical arena.

- Can you identify other governments in the Global South that, like Brazil, have mobilized renewable energy sources in their quest for strategic partnerships with state actors in the Global North?
- The idea of a transnational electricity supergrid for Asia has been much publicized in recent years. Now, what about the corresponding prospects for *natural gas*? To what extent have foreign policymakers in the region responded to the Saint-Simonian imperative in this case? What are the similarities and differences compared to the European and South American experiences in regional gas integration discussed in this chapter?

Notes

1 Svedberg, "Saint-Simon's Vision of a United Europe", 154–158; Högselius et al., *Europe's Infrastructure Transition*, 24, 27, 60.
2 Schönholzer, "Ein elektrowirtschaftliches Programm für Europa", 385.
3 Gall, "Atlantropa".
4 Lagendijk and Van der Vleuten, "Inventing Electrical Europe".
5 Lagendijk, *Electrifying Europe*, 117.
6 Ibid., 165–167.
7 Lagendijk and Van der Vleuten, "Inventing Electrical Europe".
8 For a detailed discussion, see Lagendijk, *Electrifying Europe*, 144ff; Lagendijk and Van der Vleuten, "Inventing Electrical Europe".
9 Kaijser, "Transborder Integration of Electricity and Gas", 7–10.
10 Lagendijk, *Electrifying Europe*, 151–152.
11 Högselius, "Connecting East and West?", 249ff; Holmberg, *Survival of the Unfit*.
12 Sistemnyi operator edinoi energeticheskoi sistemy, "Istoriya".
13 Tchalakov et al., "Bulgarian Power Relations"; Hegmann, "Die Entwicklung der Zusammenarbeit im RGW", 21.
14 Sebitosi and Okou, "Rethinking the Power Transmission Model for Sub-Saharan Africa", 1448–1451.
15 Gutierrez Ramirez, "Energy Integration".
16 "Declaration of 9th May 1950 delivered by Robert Schuman", *European Issue*, no. 204, 10 May 2011.
17 Alter and Steinberg, "The Theory and Reality of the European Coal and Steel Community", 6.
18 Högselius et al., *Europe's Infrastructure Transition*, 82–83.
19 Ibid., 83.
20 Högselius et al., "Natural Gas in Cold War Europe".
21 Kaijser, "Striking Bonanza".
22 Högselius *Red Gas*, Chapter 7.
23 Mares and Martin, "Regional Energy Integration in Latin America".
24 Ibid.; Kellogg, "Regional Integration in Latin America", 192; "Colombia, Venezuela to Build 2,000-Mile Oil Pipeline". *Colombia Reports*, 25 October 2011.
25 Mares and Martin, "Regional Energy Integration in Latin America", 57.
26 Ibid.
27 See, for example, "Argentina Gradually Organizing Natural Gas Production and Consumption Priorities", *MercoPress*, 16 March 2018; "Natural Gas to Become Prime Source of Energy in Latin America", *OilPrice.com*, 27 February 2018.

28 Eran et al. "The Gas Deal with Egypt".
29 Högselius, *Red Gas*, 47.
30 See, for example, Nilsson and Filimonova, "Russian Interests in Oil and Gas Resources in the Barents Sea".
31 Fischer, *History of the International Atomic Energy Agency*, 20–21.
32 Ibid., 29.
33 Ibid., 60–61.
34 Ibid., 61–62.
35 Ibid., 63.
36 Quoted in Schrafstetter and Twigge, "Spinning into Europe", 260.
37 Högselius, "The Decay of Communism".
38 Högselius et al., *Europe's Infrastructure Transition*, 93.
39 Hecht, "The Power of Nuclear Things", 10–20.
40 For a theoretical discussion of these terms, see, for example, Loomba, *Colonialism/Postcolonialism*, 11ff. "Informal colonialism" is sometimes used as an alternative concept.
41 Högselius, "Connecting East and West?"
42 There is now an initiative, championed by the Baltics and the EU, to enable synchronization of the Baltic grid with the West European grid by 2025. This would then require delinking of the Baltics from Russia and Belarus. It remains to be seen whether this project materializes. See, for example, "EU to Work with Baltic States on Decoupling from Russian Power Grid", *Reuters,* 1 June 2017.
43 Mathews and Tan, "Manufacture Renewables to Build Energy Security", 167.
44 Mathews, "Biofuels", 3351.
45 Lorenzo and Vazquez, "The Rise of Biofuels in IR", 909–910.
46 Mano et al., "Gobitech and Asian Super Grid for Renewable Energies in Northeast Asia", 14; "How Asia's Super Grid may open a brighter future for the region", *Global Construction Review*, 28 June 2017.
47 United Nations, *Multi Dimensional Issues in International Electric Power Grid Interconnections*, 107.

Further reading

Fischer, David. *History of the International Atomic Energy Agency: The First Forty Years.* Vienna: IAEA, 1997.
Högselius, Per, Arne Kaijser and Erik van der Vleuten. *Europe's Infrastructure Transition: Economy, War, Nature.* Basingstoke and New York: Palgrave Macmillan, 2016.
Mares, David, and Jeremy Martin. "Regional Energy Integration in Latin America: Lessons from Chile's Experience with Natural Gas". *Third World Quarterly* 33, 1 (2012): 55–70.

Bibliography

Alter, Karen J., and David Steinberg. "The Theory and Reality of the European Coal and Steel Community". Buffett Center for International and Comparative Studies, Working Paper No. 07-001, January 2007.
Eran, Oden, Elai Rettig and Ofir Winter. "The Gas Deal with Egypt: Israel Deepens its Anchor in the Eastern Mediterranean". *INSS Insight,* No. 1033, 12 March 2018.
Fischer, David. *History of the International Atomic Energy Agency: The First Forty Years.* Vienna: IAEA, 1997.

Gall, Alexander. "Atlantropa: A Technological Vision of a United Europe." In *Networking Europe: Transnational Infrastructures and the Shaping of Europe, 1850–2000*, edited by Erik van der Vleuten and Arne Kaijser, 99–128. Sagamore Beach, MA: Science history publications, 2006.

Gutierrez Ramirez, Javier. "Energy Integration: The Central American Experience in Designing and Implementing the Regional Electricity Market". Report to the OECD, www.oecd.org/aidfortrade/casestories/casestories-2017 (accessed 2 May 2018).

Hecht, Gabrielle. "The Power of Nuclear Things". *Technology & Culture* 51, 1 (2010): 1–30.

Hegmann, Margot. "Die Entwicklung der Zusammenarbeit im RGW." *Zeitschrift für Geschichtswissenschaft* 19 (1971): 15–53.

Högselius, Per. "Connecting East and West? Electricity Systems in the Baltic Region". In *Networking Europe: Transnational Infrastructures and the Shaping of Europe, 1850–2000*, edited by Erik van der Vleuten and Arne Kaijser, 245–277. Sagamore Beach, MA: Science History Publications, 2006.

Högselius, Per. "The Decay of Communism: Managing Spent Nuclear Fuel in the Soviet Union". *Risk, Hazards and Crisis in Public Policy* 1, 4 (2010)

Högselius, Per, Anna Åberg and Arne Kaijser. "Natural Gas in Cold War Europe: The Making of a Critical Infrastructure". In *The Making of Europe's Critical Infrastructure: Common Connections and Shared Vulnerabilities*, edited by Per Högselius, Anique Hommels, Arne Kaijser and Erik van der Vleuten, 27–61. Basingstoke and New York: Palgrave Macmillan, 2013.

Högselius, Per, Arne Kaijser and Erik van der Vleuten. *Europe's Infrastructure Transition: Economy, War, Nature*. Basingstoke and New York: Palgrave Macmillan, 2016.

Holmberg, Rurik. "Survival of the Unfit: Path-Dependence and the Estonian Oil Shale Industry". PhD thesis, Linköping University, 2008.

Kaijser, Arne. "Trans-Border Integration of Electricity and Gas in the Nordic Countries, 1915–1992." *Polhem: Tidskrift för teknikhistoria* 15 (1997): 4–43.

Kaijser, Arne. "Striking Bonanza: The Establishment of a Natural Gas Regime in the Netherlands". In *Governing Large Technical Systems*, edited by Olivier Coutard, 38–57. London: Routledge, 1999.

Kellogg, Paul. "Regional Integration in Latin America: Dawn of an Alternative to Neoliberalism?" *New Political Science* 29, 2 (2007): 187–209.

Lagendijk, Vincent. *Electrifying Europe: The Power of Europe in the Construction of Electricity Networks*. Amsterdam: Aksant, 2008.

Loomba, Ania. *Colonialism/Postcolonialism*. Second edition. London and New York: Routledge, 2005.

Lorenzo, Cristian, and Patricio Yamin Vazquez. "The Rise of Biofuels in IR: The Case of Brazilian Foreign Policy Towards the EU". *Third World Quarterly* 37, 5 (2016): 902–916.

Mares, David, and Jeremy Martin. "Regional Energy Integration in Latin America: Lessons from Chile's Experience with Natural Gas". *Third World Quarterly* 33, 1 (2012): 55–70.

Mathews, John. "Biofuels: What a Biopact between North and South Could Achieve." *Energy Policy* 35 (2007): 3550–3570.

Mathews, John, and Hao Tan, "Manufacture Renewables to Build Energy Security". *Nature* 513 (11 September 2014): 166–168.

Mano, Shuta, Bavuudorj Ovgor, Zafar Samadov, Martin Prudlik et al. "Gobitech and Asian Super Grid for Renewable Energies in Northeast Asia". Energy Charter Secretariat and the Energy Economics Institute of the Republic of Korea, 2014.

Nilsson, Annika E., and Nadezhda Filimonova. "Russian Interests in Oil and Gas Resources in the Barents Sea". Stockholm Environmental Institute, Working Paper 2013-05, 2013.

Schönholzer, Ernst. "Ein elektrowirtschaftliches Programm für Europa." *Schweizerische Technische Zeitschrift* 23 (1930): 385–397.

Schrafstetter, Susanna, and Stephen Twigge, "Spinning into Europe: Britain, West Germany and the Netherlands – Uranium Enrichment and the Development of the Gas Centrifuge, 1964–1970". *Contemporary European History* 11, 2 (2002): 253–272.

Sebitosi, A.B. and R. Okou. "Rethinking the Power Transmission Model for Sub-Saharan Africa". *Energy Policy* 38 (2010): 1448–1454.

Sistemnyi operator edinoi energeticheskoi sistemy. "Istoriya". www.so-ups.ru/index.php?id=925 (accessed 15 May 2014).

Sörgel, Herman. *Die drei großen "A": Großdeutschland und italienisches Imperium, die Pfeiler Atlantropas.* München: Piloty & Loehle, 1938.

Svedberg, The. "Saint-Simon's Vision of a United Europe". *European Journal of Sociology* 35, 1 (1994): 145–169.

Tchalakov, Ivan, Tihomir Mitev and Ivaylo Hristov. "Bulgarian Power Relations: The Making of a Balkan Power Hub". In *The Making of Europe's Critical Infrastructures: Common Connections and Shared Vulnerabilities*, edited by Per Högselius, Anique Hommels, Arne Kaijser and Erik van der Vleuten, 131–156. Basingstoke and New York: Palgrave Macmillan, 2013.

United Nations. *Multi Dimensional Issues in International Electric Power Grid Interconnections*. New York: United Nations, 2006.

Index

Note: Page numbers in italics refer to 'figures'.